Aelfric's
De Temporibus Anni

EARLY ENGLISH TEXT SOCIETY

Original Series No. 213

1942 (for 1940), reprinted 1970

End of Ælfric's *Catholic Homilies*, *Oratio*, and beginning of *De Temporibus Anni*.
Cambridge University Library MS. Gg. 3. 28, fol. 255r.

Aelfric's
De Temporibus Anni

EDITED FROM ALL THE KNOWN MSS.
AND FRAGMENTS
WITH AN INTRODUCTION
SOURCES, PARALLELS, AND NOTES

BY

HEINRICH HENEL

Published for
THE EARLY ENGLISH TEXT SOCIETY

by

OXFORD UNIVERSITY PRESS
LONDON   NEW YORK   TORONTO

OXFORD
UNIVERSITY PRESS

Great Clarendon Street, Oxford OX2 6DP
United Kingdom

Oxford University Press is a department of the University of Oxford.
It furthers the University's objective of excellence in research, scholarship,
and education by publishing worldwide. Oxford is a registered trade mark of
Oxford University Press in the UK and in certain other countries

First Edition published in 1942
Reprinted 1970

Published in the United States of America by Oxford University Press
198 Madison Avenue, New York, NY 10016, United States of America

British Library Cataloguing in Publication Data
Data available

Library of Congress Cataloging in Publication Data
Data available

Original Series, 213

ISBN 978-1-84-384412-9

TO
MAX FÖRSTER
IN GRATITUDE AND
AFFECTION

PREFACE

A BOOK should speak for itself and not stand in need of lengthy explanations. I would ask only that the reader should consider this edition as a whole, and keep in mind the fact that its parts are interrelated and meant to be used together. The third section of the Introduction, for instance, cannot be read intelligently unless reference is made to the text. The latter, in its turn, is based on the conclusions reached in the Introduction and, occasionally, in the notes. The sources especially are often supplemented or explained in the notes. The term 'source' itself changes its meaning almost from paragraph to paragraph, the relation of Ælfric's text to its sources being discussed in the relevant section of the Introduction and in a number of notes. The notes are not meant to serve as a complete commentary; they do not repeat what can be found in any encyclopaedia or science primer, but concentrate on those passages which offer some difficulty, or which are of special interest to students of theology, the history of science, or folk-lore.

It is a pleasure to record my indebtedness to those who have assisted in the preparation of this edition. To the Master and Fellows of Corpus Christi College, Cambridge, my thanks are due for the permission to use and print from their manuscript no. 367. The frontispiece was reproduced with the kind consent of the Syndics of Cambridge University Library. The officers of several libraries in England and Germany have taken considerable trouble in providing me with books not readily accessible. I am especially obliged to Mr. Francis Wormald of the Department of Manuscripts of the British Museum, whose wide knowledge is matched only by the kindness with which he has answered my inquiries. Professor Max Förster of Munich has been my constant guide and adviser. In addition to all that he has done for the book in the course of its preparation, he has read proofs and added immeasurably to its value by his keen criticism. Finally, I wish to tender my thanks to the Society for undertaking to publish the book, and to its Secretary, Dr. Mabel Day, for the generous way in which she has given of her time in getting my manuscript ready for the press, and for her valuable suggestions and corrections.

HEINRICH HENEL

QUEEN'S UNIVERSITY,
KINGSTON, CANADA,
July, 1938

Publication of this book has been greatly delayed by the outbreak of war. The manuscript was delivered to the Society in July, 1938, and only minor additions and corrections have been possible since then. Professor Förster was able to read proofs of the Introduction only. I have, however, had the advice of my colleague Dr. A. A. Day on some points in the Latin.

The emendation on page 23, note 14, was suggested by the Reader of the Oxford University Press. Professor Laistner of Cornell University tells me that the reading *redduntur* is supported by MSS. St. Gall 255 and Paris, Bibl. Nat. lat. 12271.

December, 1940   H. H.

CONTENTS

INTRODUCTION

§ 1. THE MANUSCRIPTS

THE manuscripts used for this edition are named as follows:

(1) A = British Museum, MS. Cotton Tiberius A. III.[1]
(2) B = British Museum, MS. Cotton Tiberius B.V.
(3) C = Corpus Christi College, Cambridge, MS. 367.
(4) D = British Museum, MS. Cotton Titus D. XXVII.
(5) E = British Museum, MS. Cotton Caligula A. XV. *First Text.*
(6) F = British Museum, MS. Cotton Caligula A. XV. *Second Text.*
(7) G = University Library, Cambridge, MS. Gg. 3. 28.
(8) H = Vatican Library, MS. Reginense Lat. 1283.

We have, then, knowledge of eight MSS. in all.[2] E and H offer but short fragments; F contains merely eight of the total of fourteen chapters, and C lacks the first chapter. There remain four MSS. that have all fourteen chapters, but of these again G alone has the complete text, the other three being deficient to a greater or lesser degree.

G is outstanding also in that it contains nothing but Ælfrician texts. Here alone *De Temporibus Anni* is found among other works of its author.[3] It was included in this

[1] A copy of MS. A is preserved in the Bodleian Library at Oxford, MS. Junius 41.
[2] R. Wülcker, *Grundriss zur Geschichte der angelsächsischen Litteratur*, Leipzig, 1885, p. 477 sq., lists a ninth MS., Cotton Otho A. XV, which he says was but a fragment even in Wanley's time and was destroyed in the fire of 1731. He refers to Wanley, p. 234. There never was a MS. Otho A. XV; at least Wanley nowhere mentions it. On p. 234 Wanley describes Caligula A. XV, which indeed contains a fragment (actually two fragments) of *De Temporibus Anni*. Immediately preceding Caligula, Wanley describes MSS. Otho A. VIII, X, XII, and XIII, and following Caligula A. XV he describes Otho A. XVIII. Thus Wülcker's error is obvious. He misread Caligula A. XV for Otho A. XV and in this manner introduced a MS. which never existed.
[3] MS. A contains four works of Ælfric, but they are short and not written together, so that the book cannot be called an Ælfrician codex.

codex because it was, or was supposed to be, Ælfric's. MSS.
ABDE offer our tract as an anonymous treatise on science,
mixed up with other 'scientific' matters, prognostics and computus rules and tables. In MSS. C and F we find *De Temporibus Anni* on loose sheets of whose provenance nothing is
known and which are bound up with MSS. of different origin.
We may say, then, that in MS. G alone the tradition of *De
Temporibus Anni* is 'literary', whereas it is 'scientific' in all
other MSS. that can be judged in this connexion.

(1) *British Museum, MS. Cotton Tiberius A. III.* (*A*)
Descriptions:
   H. Wanley in George Hickes's *Thesaurus*, Oxford, 1705,
   ii. 193–9.
   J. Planta, *A Catalogue of the Manuscripts in the Cottonian
   Library*, London, 1802, p. 31 sq.
   H. Logeman, *Anglo-Saxon and Latin Rule of St. Benet*,
   E.E.T.S., O.S. 90 (1888), pp. xx–xxv.
   Max Förster, *Archiv für das Studium der neueren
   Sprachen und Literaturen*, cxxi (1908), pp. 31–45.
   Max Förster, *Englische Studien*, lx (1925), p. 66 sq.
   (supplements the description in *Archiv*, cxxi).

Date. Wanley says that the MS. is of pre-Conquest date.
Förster in his most detailed and exact description of the codex
distinguishes six parts of separate provenance, bound together, no doubt, at the order of Sir Robert Cotton. He finds
that at least ten different hands are responsible for the writing, most of them belonging to the first half or the middle of
the eleventh century.[1] *De Temporibus Anni* is found in the
second (fols. 57–116) of the six MSS. that now make up the
book. In addition, three other writings of Ælfric are copied,
whole or in part, in this second MS. The four texts together
occupy about one-third of the pages of the MS. Förster thinks
it was written about the middle of the eleventh century.

---

[1] In *Englische Studien*, lx. 66, Förster attributes the writing of the first
MS. to the end of the eleventh century, while Fehr, *Die Hirtenbriefe Ælfrics*,
Greins Bibliothek, vol. ix, p. xvii, says it was written about 1100.

Provenance. The same authority assigns the origin of this second part to the south-east. Fehr, loc. cit., says 'the language shows clearly a Kentish influence'. W. S. Logeman[1] believes that the first part of MS. Tiberius A. III has 'the most intimate relation' with an ancient codex from Christ Church, Canterbury, whilst Förster[2] is satisfied that it is identical with the ancient codex described in the Christ Church Catalogue of MS. Cotton Galba E. 4.[3] The second part of the MS. also would seem to have come from Canterbury. This is indicated not only by the Kentish character of the language but also by the Litany on fols. 112*v*–113*r*. In it, the names of three saints are capitalized, Margaret, Augustine, and Dunstan. Amongst the others is St. Mildred, of whom Alban Butler[4] says: 'In 1033, the remains of St. Mildred were translated to the monastery of St. Austin's at Canterbury, and venerated above all the relics of that holy place, says Malmesbury'. It might be mentioned in passing that the Old English accounts of the monastery of St. Mildred in Thanet of MSS. Cotton Caligula A. XIV and Lambeth 427[5] do not as yet know of the translation. The former states:[6] '*Sancte mildryð resteð binnan teneð on ðæm iglande*' ; and the latter:[7] '*Sancta eadburh þa to ðam mynstre feng æfter sancte myldrype; 7 heo ða cyricean arærde ðe hyre lichama nu inne resteð*'.[8] Another noteworthy name in the Litany is that of St. Elphege. He 'was martyred in 1012 and his relics were translated from London to Canterbury

---

[1] *Anglia*, xv. 24 sq.

[2] *Archiv*, cxxi. 31.

[3] The latter was edited by the late Dr. M. R. James in his *The Ancient Libraries of Canterbury and Dover*, Cambridge, 1903. Cp. pp. 50 and 508.

[4] *The Lives of the Fathers, Martyrs and other Principal Saints*, Dublin and London, 1838, i. 25.

[5] Printed by Cockayne, *Leechdoms*, iii. 422–33.

[6] *Leechdoms*, iii. 424.

[7] *Leechdoms*, iii. 430 and Förster, 'Die altenglischen Beigaben des Lambeth-Psalters', *Archiv*, cxxxii. 334.

[8] This sentence is also found in *Die Heiligen Englands* (ed. F. Liebermann, Hannover, 1889, p. 5). The Latin version of the *Halgan* (ibid., p. 6) has an addition in this place wherein the translation of St. Mildred to St. Augustine's, Canterbury, is expressly noted.

in 1023'.[1] We have, then, a litany which singles out St. Augustine and St. Dunstan for special honours and which includes St. Mildred and St. Elphege. The conclusion seems indicated that the MS. was written after A.D. 1033, and that it originated at St. Augustine's, Canterbury.[2] *De Temporibus Anni* is found on fols. 65*v*–73*r*. This is the poorest among the complete MSS. of our text, inferior even to C which is much later. Cockayne failed to notice the first chapter which, in this MS., is placed at the end.[3] A definite relation can be established between this MS. and Titus D. XXVI–XXVII (the latter two originally must have been one). The four texts immediately preceding *De Temporibus Anni* on fol. 65 of Tiberius A. III, viz. *De minuendo sanguine, De nativitate infantium, Lunaris Sancti Danielis,* and *Lunaris de aegris,* are repeated, in this order, in MS. Titus D. XXVI, fols. 6*r*–9*r*. A complete list of all texts that occur in both the second part of Tiberius A. III and Titus D. XXVI–XXVII is appended. It is of interest because it shows in what company the scribes put Ælfric's learned treatise, and because it indicates that the two MSS. derived part of their materials from the same source. These materials need not have been copied from the same book, but they clearly go

---

[1] Bishop and Gasquet, *The Bosworth Psalter*, London, 1908, p. 27. Cp. ibid., p. 31, and also the important remarks (p. 32 sq.) on the cults of saints at Canterbury just before and after the Conquest.

[2] The matter apparently is more difficult than I had suspected. Mr. F. Wormald writes me as follows: 'The Litany of Tiberius A. III does *not* contain definite evidence for attributing the MS. to St. Augustine's, Canterbury. Personally I am convinced that it is a litany of Christ Church, Canterbury. In the first place: None of the ancient archbishops of Canterbury are mentioned with the exception of St. Augustine, and as they were buried at St. Augustine's you would expect them to find a place there. Another thing is the position of St. Mildred. If the Litany came from St. Augustine's I should have expected her to come higher up in the list, and anyway before St. Etheldreda who precedes her here. Also in favour of Christ Church is the presence of saints in the Litany whose relics are known to have been at Christ Church. They are SS. Ælfeage, Salvius, Blasius, Dunstan, Audoenus, Swithin, Furseus, and Astroberhta. Salvius and Furseus are particularly indicative, though both occur in the Bosworth Psalter which I am pretty well convinced is St. Augustine's, in spite of Edmund Bishop's remarks to the contrary. On the other hand there are none of the archbishops.'

[3] Cp. Förster, *Archiv*, cxxi. 40.

back to a common archetype. For Tiberius A. III I follow the numbering of Förster, and for Titus D. XXVI that of W. de Gray Birch.[1] Birch's description of Titus D. XXVII is so inadequate that I quote the pages rather than his numbers.

Titus D. XXVI.	*Texts.*	*Tiberius A. III.*
2	*The Ages of the World*	21
7	*De minuendo sanguine*	29
8	*De nativitate infantium*	30
9	*Lunaris Sancti Danielis de nativitate*	31 (and 6)
10	*Lunaris de aegris*	32 (and 7)
11	*Lunaris de somnis*	4
12	*De tonitruis dierum*	8
13	*Signa de temporibus*[2]	5 (and 17)
14	*De somniorum diversitate*[3]	2

Titus D. XXVII.

fol. 27r–29v	*De observatione lunae*[3]	3
fol. 30r–54r	*De temporibus anni*	33

(2) *British Museum, MS. Cotton Tiberius B. V (B).*

Descriptions:

H. Wanley in George Hickes's *Thesaurus*, Oxford, 1705, ii. 215–17.

J. Planta, *A Catalogue of the Manuscripts in the Cottonian Library*, London, 1802, p. 35 sq.

J. A. Herbert, *Illuminated Manuscripts*, London, 1911, p. 113 sq.

M. R. James, *Marvels of the East*, Oxford, 1929, pp. 2–6.

Date. Herbert and James agree that this is a pre-Conquest, eleventh-century MS. Fehr[4] erroneously gives the date as A.D. 969, mistaking the opening year of the Easter table

---

[1] In his 'On Two Anglo-Saxon Manuscripts in the British Museum', *Transactions of the Royal Society of Literature*, Second Series, vol. xi, London, 1878, pp. 463–512. He described the two MSS. Titus D. XXVI–XXVII again on pp. 251–83 of *Liber Vitae, Register and Martyrology of New Minster and Hyde Abbey*, London and Winchester, 1892.

[2] Titus D. XXVII, fol. 25 is similar, but not the same.

[3] The texts in Titus and Tiberius are similar, not identical. For *De observatione lunae* see Emanuel Svenberg, *De Latinska Lunaria*, Göteborg, 1936, pp. 12 sq., 18, 21.

[4] *Texte und Forschungen zur englischen Kulturgeschichte, Festgabe für Felix Liebermann*, Halle, 1921, p. 32.

fol. 16*r* for the date of the MS. itself. Elsewhere[1] I have pointed out the dangers inherent in dating MSS. by their Easter tables. The *terminus a quo* for this MS. is the year 993, when Ælfric's *De Temporibus Anni* was probably written. A *terminus ad quem* might be found in the list of West Saxon kings on fol. 22*r* which ends with Æthelred.[2]

[1] *Studien zum altenglischen Computus*, Leipzig, 1934, p. 23.
[2] Printed by Thomas Wright, *Reliquiae Antiquae*, London, 1845, ii. 171. A critical edition of the lists of bishops and kings in MS. Tiberius B. V would be most desirable. I note below the dates of the last bishop given for each diocese. These dates are taken from William Stubbs, *Registrum Sacrum Anglicanum*, Oxford, 1897; W. G. Searle, *Anglo-Saxon Bishops, Kings and Nobles*, Cambridge, 1899; Oilard and Crosse, *Dictionary of English Church History*, Oxford, 1912. Where more than one date is given for accession or death, the dates are in doubt.

Canterbury:	Sigeric	989 : 990–994 : 995
Rochester:	Ælfstan	946 : 964–995
London:	Ælfstan	961–995 : 996
Selsey:	Ordbeorht	989–1009
Winchester:	Ælfheah II	984–1005 (to Canterbury)
Sherborne:	Æthelsige I	978–990 : 992
Ramsbury:	Sigeric	985–990 (to Canterbury)
Wells:	Sigegar	975–995 : 997
Crediton:	Ælfweald II	985 : 988–988 : 1008
Worcester:	Heathured	781–798 : 800
Lichfield:	Cynefrith	826 : 836–841 : 845
[Hereford]:	Eadwulf	825 : 832–836 : 839

N.B. There is no title to this item. It contains, however, the bishops of Hereford.

[Dorchester]:	Æscwig	975 : 979–1002

N.B. The MS. text describes this item as the list of Lindisfarne, but in actual fact the first nine names given are those of bishops of Lindsey, and the last three are bishops of Dorchester.

Elmham:	Theodred I	964 : 974–979 : 982
York:	Wigmund	837–854
Hexham:	Eanbeorht	800–813
Lindisfarne:	Ecgbeorht	802–821
Whithern (Casa Candida):	Beaduwulf	791–802
[Glastonbury]:	Sigegar	965–975 (to Wells)

N.B. The last list, following the genealogies of the English kings, is again without a title, but it has the names of the abbots of Glastonbury.

From the dates given it seems likely that the lists were originally compiled in the ninth century, and that they were continued, about the year 989, by a person who lived in the south and had no information about the bishops of the northern dioceses. The continuator gives the name of Sigeric (he is the

For the earlier kings the length of rule is noted, the number of years in all cases, sometimes also the extra months and weeks. There is no such note with the name of Æthelred (†1016), which may permit the conclusion that the king was still alive when the MS. was written. It is likely, however, that the lists of bishops and kings in this MS. were drawn up as early as 989 (see p. xiv, note 2) and that we have here merely a copy, not the original. The number of mistakes and omissions makes this almost a certainty.

Provenance. At one time this codex belonged to Battle Abbey.[1] It may have come there from Exeter, as is suggested by Förster.[2] There is no doubt that the computus of MS. Tiberius B.V., i.e. fols. *2r–19r*, shows the closest resemblance with the computus of another Exeter book, viz. the Leofric Missal B, i.e. its Anglo-Saxon part.[3] On the other hand, some weight may attach to the fact that on fol. *23v* our MS. notes the cities which 'our bishop Sigeric' saw on his journey to Rome. Sigeric was archbishop of Canterbury A.D. 989–95, and his journey is noted under date of A.D. 990 in the annals of MS. Caligula A. XV, written by a monk of Christ Church, Canterbury.[4] To him are addressed the Latin prefaces of both parts of Ælfric's Homilies. It is conceivable, therefore, that the entry regarding his journey indicates some sort of connexion of MS. B with Canterbury.

It might also be mentioned that the texts offered by Tiberius B. V on fols. *30r–54v* are, with but minor exceptions, identical with the contents of MS. Tiberius C. I, fols. *19a–42b*. Most of the first part of the present codex (i.e. up to fol. 88) belonged at one time to John Lord Lumley.[5] It

---

only one to whom an epithet is given: *dei amicus*) both as the last bishop of Ramsbury and as the last archbishop of Canterbury. This points to the year 989, when Sigeric was archbishop-designate.

[1] James, *Marvels of the East*, p. 6.
[2] *The Exeter Book of Old English Poetry*, London, 1933, p. 13, note 6. Cp. also ibid., p. 49, note 31a.
[3] Cp. *Studien zum altenglischen Computus*, pp. v, 13, 14, note 41, 19, note 57, 22, 29. [4] See below, p. xxii.
[5] See British Museum MS. Additional 36659, fols. 212, 213.

seems possible to me that a different MS. begins also with
fol. 30. The writing is larger, the content agrees with a different set of MSS. than that of the first twenty-nine fols., and
two leaves are missing before fol. 30. Certainty cannot be
gained, however, for the leaves are inlaid, and thus the
original structure of the volume is destroyed.
De Temporibus Anni occupies fols. 24r–28v (or fols. 23–7
of the old pagination). Next to G, this MS. is the best of our
text. Probably as old as G, it is almost as complete, but not
as correct or reliable. This is not surprising since G contains
nothing but Ælfrician texts, whereas B offers a compilation
of computus matters, prognostics, science, and odd learning.
G may derive directly from the autograph, but B is separated
from it by at least one intermediary link.

K. Sisam assumes that *De Temporibus Anni* was entered,
at the author's direction, in a MS. which already contained
his Homilies and from which the scribe of G took his copy.
He further thinks there are 'indications that Ælfric retouched
the version in Tiberius B. V at beginning and end'[1] in order to
divorce it from the Homilies and to make it fit for separate
issue. I agree with the first assumption but cannot see that
the second is supported by sufficient proof. *Pluccian* (B) for
gadrian (all other MSS.) in the first sentence of the tract is
indeed an improvement. However, B retains the introductory phrase *Ic wolde eac* which is meaningless unless it refers
to a preceding text. It alone of all MSS. repeats part of the
prefatory sentence (placing it *after* the first paragraph) which
is found in G and which says that what follows is *not* a sermon. Finally, B replaces the Latin *Explicit* of G (missing in
all other MSS.) by a short colophon, *god helpe minum
handum*.[2] One would think that Ælfric, had he made these

---

[1] *Review of English Studies*, vol. viii (1932), p. 52, note 4.
[2] A similar addition is found at the end of MS. D. It cannot be Ælfric's
both because of the date of D and because it is attached to the first chapter,
which stands last in this MS. Cp. Charles Plummer, 'Colophons and
Marginalia of Irish Scribes', *Proceedings of the British Academy*, London,
1926. Why the scribe uses the plural *handum* is explained by M. Förster,
Archiv für das Studium der neueren Sprachen, clxii. 230.

Ælfric's *De Temporibus Anni*, Chapters IV. 52–VI. 11.
Corpus Christi College, Cambridge, MS. 367, Second Part, fol. 7r.

changes, would have made a better job of them, i.e. he would have omitted all reference to the Homilies. There is nothing in the changes of B that could not be the work of a scribe.

(3) *Corpus Christi College, Cambridge, MS. 367.* (C)
Descriptions:
    M. R. James, *A Descriptive Catalogue of the Manuscripts in the Library of Corpus Christi College, Cambridge,* Cambridge, 1912, ii. 199–204.
    A. S. Napier, 'Leofric's Vision', *Transactions of the Philological Society,* 1908, pp. 180–8.

Date. According to James, the covers of the present book enclose 'five volumes of cent. XV, XI–XII, XIV, [XIII, and] XI, respectively'. The first part does not concern us here; it is a paper MS. of the fifteenth century. With the second part pagination begins afresh. It contains *De Temporibus Anni* and a number of fragments of homilies in Anglo-Saxon. The leaves are bound up in a very confused fashion. *De Temporibus* is found on fols. 1, 2, 7–10*r*, while fragments from Ælfric's Homilies occupy both the intervening and following pages. The six leaves of *De Temporibus* are from a separate book; they have from 30 to 39 lines to a page, whereas the Homilies have 30 (from fol. 17 only 27) lines to a page. The writing of *De Temporibus* is excellent, but smaller than that of the surrounding pages.

James, p. 200, says that the writing of *De Temporibus* is 'in a smallish hand of cent. XI–XII' and describes that of the Homilies as 'a good hand of cent. XII'. One hesitates to contradict an authority such as the late Dr. James, but I am inclined to think that the writing of *De Temporibus* is, if anything, later than that of the Homilies. I should suggest the early twelfth century as its date.

The facsimile plate to face will show that the Caroline minuscule *a* is used almost exclusively rather than the insular *a*. For *æ* we find the insular ligature, but occasionally the continental *a* with an upward stroke at the right. The low insular *f* preponderates, but the tall continental character is not uncommon. The ratio on fol. 1*v* is 28:7 in favour of the former. The low insular *r* is used sixty-four times on fol. 1*r* as against six occurrences of the Caroline *r*. The low insular *s* is not used at all, the tall continental *s* taking its place in the majority of cases. Round *s* is not infrequent, either. The straight Caroline *d* does not occur, nor the deeply bent *d* of insular script. The character used is a *d* with a pretty long neck slightly bent to the left. There is hardly any difference between *d* and *ð*, excepting the stroke across the neck. Two types of the letter *y* seem to be used indiscriminately, with either a long left or a long right leg. The two types occur close beside each other in the first line of fol. 2*v*. The letter *u* often has a short tail bent back to the left. Several times we find the Latin contractions for *ri* and *de* that are not commonly met in Anglo-Saxon texts. The sign *þ̵* is used repeatedly by mistake for *þ*. The scribe cannot have known much Anglo-Saxon. Erasures, corrections, and omissions that destroy the meaning (apart from the omission of whole sentences) are frequent. He has obvious difficulty with the insular letters, writing *ontenð* for *ontend* (iii. 11), *odrum* for *oðrum* (viii. 15 and x. 23), *awplum* for *applum* (iv. 53), *nipe* for *niwe* (viii. 6).

All this might not prove more than that the MS. is of post-Conquest date, and that it was written by a scribe who was accustomed to the continental, rather than to the insular, characters. However, the angularity of the individual letters and the crowding of the lines are heralds of a later age. The early twelfth century is, I think, the right date of this hand.

Provenance. A letter by the Abbot of Westminster to the Prior of Worcester is found on fol. 52*r* of our part of the MS. According to James it must have been written between 1130 and 1140. Since, however, the six leaves that contain *De Temporibus* are of separate origin, there is no evidence that they, too, had some connexion with Worcester. There is no clue to their place of origin. The three sheets must have been kept apart, and unbound, for some time, as fols. 1*r*, 2*v*, 7*r*, 8*v*, 9*r*, and 10*v* are yellowed quite markedly.

De Temporibus Anni. Its place in the MS. has already been noted. The copy is bad and practically useless for the reconstruction of the text. It was not used by Cockayne or any previous editor, since the Anglo-Saxon portions of this MS. were not described by Wanley and remained unknown to students until Dr. James discovered them during his work of cataloguing. The chief value of this copy lies in the fact that it bears witness to the continued interest in Ælfric's treatise long after the Conquest, for which there is no other evidence.

A number of paragraphs or whole sentences are omitted in C. There is method in these omissions (there are also inadvertent omissions of a few words). The last sentences of chapters ii, vi, and x, which are lacking, contribute little to the substance of the treatise. Chapter iii. 27–9 deals with the reckoning of Easter and chapter vi. 2–6 with the right date for the vernal equinox. It is rather striking that these passages should be omitted, considering the interest that attached to these matters in the Middle Ages. It must be assumed that the questions here dealt with, which were still open to doubt, and gave rise to confusion, in the age of Ælfric, had been settled satisfactorily by the time the scribe of C made his copy.

(4) *British Museum, MS. Cotton Titus D. XXVII. (D)*
Descriptions:

H. Wanley in George Hickes's *Thesaurus*, Oxford, 1705, ii. 247 sq.

J. Planta, *A Catalogue of the Manuscripts in the Cottonian Library*, London, 1802, p. 567.

Ferdinand Piper, *Die Kalendarien und Martyrologien der Angelsachsen*, Berlin, 1862, p. 67 sq. (for the Calendar).

W. de Gray Birch, 'On Two Anglo-Saxon Manuscripts in the British Museum', *Transactions of the Royal Society*

of Literature, Second Series, vol. xi, London, 1878,
pp. 463–512. Also *Liber Vitae: Register and Martyrology of New Minster and Hyde Abbey*, London, 1892,
pp. 269–83.

Date. This MS. was shown in the British Museum in 1931
during the Cotton Memorial Exhibition with the following
description: 'Prayers etc. in Latin and English: part of a
volume written about 1012–20 for Aelfwin, monk of New
Minster who became Abbot in 1035.' The same date is given
by J. A. Herbert.[1] Birch first assigns the MS. to the early
years of the eleventh century, but later he attributes it to 'the
concluding years of the Saxon dynasty'.[2] Förster gives three
dates: 'between 1034 and 1057', 'around 1020', and 'between
1035 and 1052'.[3] F. Wormald[4] decides on the years A.D. 1023–
35. It would seem useless to review again the evidence for
the various dates suggested. Suffice it to say that Titus is
certainly a pre-Conquest, eleventh-century MS. Fehr[5] is
definitely wrong in giving the date as 978.

Provenance. The book can be assigned without any doubt
to the New Minster at Winchester. Its computus bears the
closest resemblance with those of two other Winchester
MSS., Corpus 422 and Arundel 60, and is related to those in
three more Winchester books, Vitellius E. XVIII, Trinity
College, Cambridge, R 15. 32, and Arundel 60.[6] It has long
been known that MSS. Titus D. XXVI and XXVII originally
formed one book, and that they were separated only by Sir
Robert Cotton's binder. What does not seem to have been
noticed so far is that XXVII represents the first portion of
the ancient book. It should, therefore, be placed before
XXVI if the two volumes were ever to be united again under

[1] *Illuminated Manuscripts*, London, 1911, p. 117.
[2] Op. cit., pp. 494, 501.
[3] The first two dates in *Archiv*, cxxi. 33, and the last in *Englische Studien*, lx. 62.
[4] *English Kalendars before A.D. 1100*, London, 1934, p. 113.
[5] *Texte und Forschungen . . . für Felix Liebermann*, p. 32, following Wanley, p. 248. Here again the Easter table gave rise to the erroneous dating.
[6] Cp. *Studien zum altenglischen Computus*, pp. 13 sq. and 19.

one cover. Proof is found in the fact that computus and calendar occupy the first few pages of XXVII, and that these are always placed at the beginning in ancient liturgical MSS. *De Temporibus Anni* is written on fols. 30*r*–54*r*. MS. D ranges after G and B in reliability. It is superior to A, C, and F. As in MS. A, the first chapter is placed at the end. In D, however, there follows a short colophon, *þam sy wuldor and lof mid fæder and halgan gaste on ealra worulda woruld a butan ende. amen.* This led Birch[1] to believe that the end of chapter i was Ælfric's 'peroration'. In actual fact, however, the colophon is the scribe's, modelled on the end of the second volume of Ælfric's Homilies.[2] The Homilies must have been known well at Winchester,[3] where MS. D was written. The phrase *ða geleaffullan on Godes gelaðunge* in *De Temporibus Anni*, i. 37 brought to the scribe's mind the last sentence of the Homilies, which reads: *Nu bidde we ðone Ælmihtigan Hælend þæt he us . . . gelæde to ðære ecan gelaðunge heofenan rices, on ðam þe he rixað . . . mid his Ælmihtigan Fæder and þam Halgan Gaste, on ealra worulda woruld. Amen.*

(5 and 6) *British Museum, MS. Cotton Caligula A. XV.*

(*E and F*)

Descriptions:

H. Wanley in George Hickes's *Thesaurus*, Oxford, 1705, ii. 233–4.

J. Planta, *A Catalogue of the Manuscripts in the Cottonian Library*, London, 1802, p. 45 sq.

Felix Liebermann, *Ungedruckte Anglo-Normannische Geschichtsquellen*, Strassburg, 1879, pp. 1 sqq.

Max Förster, *Archiv für das Studium der neueren Sprachen und Literaturen*, cxxi (1908), p. 33.

Max Förster, *Englische Studien*, lx (1925), pp. 74–6.

This volume is made up of three, or perhaps four, different MSS. or fragments of such. It is permanently exhibited in

---

[1] Op. cit., p. 507.   [2] Ed. Thorpe, ii. 594.
[3] Most of them, perhaps all, were written by Ælfric at Winchester.

the British Museum and described as follows: 'Treatises of
St. Jerome and St. Cyprian: with tracts on the paschal cycle,
etc. Latin. Written in pointed minuscules, in France, in
A.D. 743 (?). Vellum.' The description clearly was intended
to apply only to the first part of the volume, which indeed is
exclusively in Latin and which shows a continental eighth-
century hand. The second part begins with fol. 120. It is a typical Anglo-
Saxon computus such as is defined and described in my
Studien. That it lacks a calendar is not without parallel in
other computi.[1] It is, nevertheless, a very interesting and
complete specimen of a computus, containing a Paschal Table
with annals from Christ Church, Canterbury,[2] Old English
texts on the Egyptian days and similar superstitions,[3] and
on computus matters.[4] From its contents I am fairly certain
that this MS. forms the beginning of the old volume from
which it was taken, but I am inclined to think that it represents merely a part of that volume, being followed, perhaps,
by a missal or a psalter. According to Liebermann and
Förster[5] it was written in England near the end of the eleventh
century. In my *Studien*, p. 24, I thought I had found the
exact date (A.D. 1083), but have since discovered that the
Palaeographical Society[6] rightly rejected this date because
the entries on the lower part of fol. 141*r* and on fol. 141*v* are
by a later hand than the rest. Since the bulk of the matter
on fols. 120–41 agrees with the other eleventh-century, pre-
Conquest computi, they should probably be assigned to the
period just preceding the Conquest. Förster,[7] following

---

[1] Cp. *Studien*, p. 4.   [2] Edited by Liebermann, op. cit., pp. 3–8.
[3] Edited variously by Cockayne, *Leechdoms*, vol. iii; by Förster, *Archiv*,
cxxi, *Englische Studien*, lx, and in *Studies in English Philology . . . in
Honour of Frederick Klaeber*; and by myself in *Englische Studien*, lxix.
[4] Edited by Cockayne, loc. cit., and in my *Studien*.
[5] *Englische Studien*, lx. 74 and *Neusprachliche Studien. Festgabe Karl
Luick* (6. Beiheft zu *Die Neueren Sprachen*), Marburg, 1925, p. 191. In the
Archiv, cxxi. 33, Förster gave the date as 'around 1000', which seems too
early.
[6] Series I, vol. iii, plate 145 (London, 1873–83).
[7] *Englische Studien*, lx. 76.

Liebermann, says that the annals (beginning fol. 132*v*) 'originally belonged to a different MS.' But quite apart from the fact that a different MS. cannot begin at the *verso*-page of a leaf, the contents, writing, parchment, and quiring[1] of fols. 120–41 make it certain that they always belonged together. There follows a single sheet, fols. 142–3. Its contents are described in detail, and partially printed, in my *Studien*, pp. 51–4. It may be from a different MS. altogether, and again it may belong to the preceding section, being the remainder of a gathering the rest of which is lost. In the latter case it must have been an inside sheet of that quire, because it begins in the middle of a sentence. But if this was so, it is striking that on its last page only nine lines are written, the rest being left blank. There is also a change of hand at fol. 142*a*. Parts from chapters vi, vii, and viii of *De Temporibus Anni* are written on fol. 142. These portions we describe as MS. E. They were overlooked by Wanley, not used by Cockayne, not mentioned by Wülcker in his *Grundriss*.[2] A. Reum,[3] K. M. Classen,[4] and M. Förster[5] confused these extracts with the more complete copy of *De Temporibus Anni* which follows in the MS.

The three extracts are not chosen in a haphazard fashion but deal with three important matters for the reckoning of Easter, viz. the right date of the vernal equinox, the bissextile year, and the so-called *saltus lunae*. The remaining texts on these four pages, some of them drawn from Byrhtferð, also deal with the Paschal computation, so that the compilation was made for a definite purpose.

[1] The gatherings are arranged in the following way: fols. 120–7 (8 leaves), fols. 128–35 (8 leaves), fols. 136–41 (6 leaves). At the beginning of each gathering an early seventeenth-century hand (perhaps Cotton's or his librarian's) has added the following letters in the left-hand bottom corner of the *recto*-page, viz. *A* on fol. 120, *B* on fol. 128, *C* on fol. 136. (I owe this information to the kindness of Mr. Francis Wormald.)

[2] Richard Wülcker, *Grundriss zur Geschichte der angelsächsischen Literatur*, Leipzig, 1885, p. 477 sq.

[3] '*De Temporibus*, Ein echtes Werk des Abtes Ælfric', *Anglia*, xi. 495.

[4] *Über das Leben und die Schriften Byrhtferðs* (Leipzig dissertation), Dresden, 1896, p. 22 sq.   [5] *Englische Studien*, lx. 76.

The third, or perhaps the fourth part of the book is formed by fols. 144–55. The vellum is different from that used for the earlier parts of the volume, the writing is larger, the ink darker. *De Temporibus Anni*, chapters iv–x and part of chapter xi, occupies fols. 144–53. The last two leaves are blank. This is our MS. F. It was identified by Wanley and used by Cockayne for his edition. A later hand wrote across the top of fol. 144*r* *Ædthelardus de Compoto.* Cp. § 5 of this Introduction.

There is no external evidence for the age and provenance of MSS. E and F.[1] The hands and language 'of the texts would seem to indicate that both belong to the pre-Conquest half of the eleventh century.

(7) *Cambridge University Library, MS. Gg. 3. 28.* (*G*)

Descriptions:

H. Wanley in George Hickes's *Thesaurus*, Oxford, 1705, ii. 153–60.

A Catalogue of the Manuscripts preserved in the Library of the University of Cambridge, vol. iii, Cambridge, 1858, pp. 71–82.

B. Fehr, *Die Hirtenbriefe Ælfrics*, Hamburg, 1914, p. xvi.[2]

K. Sisam, 'MSS. Bodley 340 and 342: Ælfric's Catholic Homilies', *Review of English Studies*, vol. viii (1932), pp. 51–6.

Date. This MS. contains none but writings of Ælfric.

---

[1] M. R. James, *The Ancient Libraries of Canterbury and Dover*, Cambridge, 1903, pp. 49 and 508, tentatively identified MS. F with no. 287 of the Christ Church, Canterbury, Catalogue compiled under Prior Henry of Eastry early in the fourteenth century. The designation in that catalogue, *Compotus Adelardi*, is strongly in favour of the identification, since a treatise on the computus is nowhere ascribed to one Adelard excepting in our MS. F. On the other hand, MS. F contains nothing but this treatise, and its last two leaves are blank, whereas the entry in the Christ Church Catalogue mentions two further items. If these were lost from the portion of the MS. which now survives it is difficult to see why two blank leaves were left between the first and second treatises.

[2] Fehr (following Napier, *Anglia*, x. 156) announced a detailed description of the MS. by Mr. Harsley. The latter, however, died and his contemplated edition of Ælfric's Homilies never appeared.

Thorpe thought it was 'probably coeval with its author',[1] and Fehr says that it 'originated in the first quarter of the eleventh century, during Ælfric's lifetime'. K. Sisam states that the hand 'would normally be assigned to the last years of the tenth or the first years of the eleventh century', and that the codex need not be very much later than 993. Fehr remarks that the MS. was written by two hands and that, for this reason, it cannot be Ælfric's autograph. Both the observation and the conclusion are taken from Thorpe. In speaking of 'the two parts of the MS.', Thorpe must have had in mind the two books of Homilies. The second hand, then, would begin on fol. 134*r*. Actually the writing here is a little smaller, more disciplined, but also more timid. Later parts, however, have so much likeness with the 'first hand' that the variations may be accidental. On the whole, the probability is, I think, in favour of *one* scribe who wrote the whole MS.[2] Thorpe also says that there are certain orthographical differences between the two parts, but in the absence of a detailed study of this question it remains uncertain what weight, if any, should be attached to his observation. It is, however, certain that the codex is not Ælfric's autograph. This is proved by a number of scribal errors which are avoided in other MSS. of the same works.

Provenance. Fehr also claims that the MS. was given by Leofric, the first bishop of Exeter (†1072), to his cathedral church as a *full spel-boc wintres and sumeres.* Such a book is indeed mentioned in the list of Leofric's gifts to his church, and the identification was first made by Wanley.[3] The best

---

[1] Benjamin Thorpe, *The Homilies of Ælfric*, London, 1843, vol. i, p. xi.

[2] Mr. Sisam, op. cit., p. 53, note 2, says: 'There is no change of hand at the beginning of the Second Series. A rounder, more upright hand, with distinct letter forms, appears in short passages here and there, e.g. ff. 225a, 226b, 240b–241b. I take this to be the hand of the miniator; cf. the minuscule headings of *De Temporibus* at f. 261b.'

[3] *Catalogus*, pp. 80 and 153. The official *Catalogue of Cambridge University Library*, p. 71, and F. E. Warren, *The Leofric Missal*, Oxford, 1883, p. xxiv, repeat Wanley's statement. Dugdale, *Monasticon Anglicanum*, London, 1655, i. 222, did not identify the two books. Thorpe and Sisam make no mention of Wanley's remark.

edition of this list is now found in the facsimile edition of the Exeter Book, where Förster[1] refutes Wanley's assumption. In addition to Förster's arguments it might be said that MS. G lacks the dedication found in several of the genuine Leofric books. The author of the official Cambridge Catalogue first noted the signature, Leo Pylkyngton, on fol. 1*r* of our MS. Sisam and Förster have drawn the conclusion that it must have come to Cambridge from Durham. There is ample documentary proof of this. Leonard Pilkington (1524?–99) and his elder brother James (1520?–76) studied at Cambridge.[2] James was elected a Fellow of St. John's College in 1539 and Leonard in 1545. James became President of his college in 1550. Being ardent supporters of Protestantism, the brothers were obliged to flee to the Continent in 1554 to escape the persecutions under Mary I. After the accession of Elizabeth they returned to Cambridge, where, in 1559, James was made Regius Professor of Theology and Master of his college. He was elected bishop of Durham in 1560 and consecrated in the following year.[3] Leonard succeeded his brother as Master of St. John's in 1561 and later followed him to Durham, where, in 1567, he was made a canon and, in 1592, treasurer of the Cathedral Chapter.

There is no evidence that Leonard gave books to the Cambridge University Library, but James is known as a donor both to it and to St. John's. This is attested by the following entries:

(1) *Donors' Book* of Cambridge University Library, handwritten about A.D. 1658, p. 17: 'Ex dono . . . Jacobi Pilkington Episcopi Dunelmensis.' There follow eighteen titles, of which our Homiliary, named in the last place, is at once the only manuscript book and the only book in Old English.

---

[1] p. 27, note 94.

[2] The article on Leonard Pilkington in *D.N.B.* suffers from several misprints in the dates it gives. I have also consulted Thomas Baker's *History of the College of St. John* and John Pilkington's *History of the Pilkington Family*. Wherever they disagree I have preferred Pilkington's dates.

[3] James Pilkington's *Collected Works* were edited by Scholefield for the Parker Society in 1842.

(2) H. R. Luard, *A Chronological List of the Graces, Documents and other Papers in the University Registry which concern The University Library*, Cambridge, 1870, p. 4, No. 49: 'Catalogue of books given, 20 by Bp. Pilkington, 40 by Abp. Parker, 73 by Sir N. Bacon. Grace Book *Δ* pp. 331b, 332a.'

(3) Charles Sayle, *Annals of Cambridge University Library 1278–1900*, Cambridge, 1916: '1574, May 16: James Pilkington, bishop of Durham, gave twenty volumes.'

(4) Lieut.-Col. John Pilkington, *History of the Pilkington Family*, 3rd edition, Liverpool, 1912, pp. 268–71: 'Will of James Pilkington, Bishop of Durham, dated 4th February 1571–2, and proved at York, 18th December, 1576, by his wife: ..."Item. I will that my books at Auckland should be given to my school in Rivington, and to poor colleges in Cambridge, and others, *by my brother Leonard*" [1]... Second Codicil to Bishop Pilkington's Will: ... "Item. That all my books (except the English ones) shall be distributed *at the discretion of my brother Leonard*,[1] whereof the old writers to the poorest libraries in Cambridge that wanted them".'

(5) Thomas Baker (†1740), *History of the College of St. John the Evangelist, Cambridge*, edited by John E. B. Mayor, Cambridge, 1869, vol. i, p. 149: 'He (James Pilkington, Tenth Master of St. John's) left several books to the College Library in number forty-five, a catalogue whereof is at the end of Vatablus' Bible, and if we may guess at his studies from his books, he was most versed in our modern Protestant divines, such as Musculus, Brentius, Bucer, Bullinger etc. Other books he gave to the public library an. 1574 in number only twenty, but to do him right, they were much the more valuable collection.' Of Leonard Pilkington, Baker says (ibid., p. 152): 'He gave or left the college seventeen books, which not coming in till the year 1594, I suppose he died about that year. These were of much the same stamp with those of his brother, or rather of a lower form, such as Aretius, Hyperius, Sadeel, etc.'

Two questions arise from this evidence. Firstly, if Bishop

[1] My italics.

Pilkington gave our Homiliary to the University Library two years before his death, how is it that it bears not his signature but that of his brother? Had it been one of the books that Leonard distributed in execution of the bishop's will, the answer would be obvious. As it is, however, no satisfactory explanation can be given, unless we assume that Leonard acted as his brother's librarian and signed the latter's books with his own name.

Secondly, how did James Pilkington acquire the Ælfric codex? It is unlikely that he found it in Durham, since no such book is mentioned in the ancient library catalogues of that church.[1] Moreover, his interests were purely theological and not antiquarian as those of his superior, Matthew Parker. The Vatablus Bible to which Baker refers is still extant in the library of St. John's College.[2] Among the forty-five titles of Pilkington's bequest there is no Old English book. Similarly the Ælfric codex is the only Old English MS. among the twenty books which Pilkington gave to the University Library. He may have acquired it at Cambridge, for the medieval libraries of the monastic colleges of Cambridge have for the most part disappeared. A more probable assumption would be that he never owned the book but had it on loan from Archbishop Parker. Pilkington's name appears with those of the two archbishops and of twelve other bishops in the extracts from Ælfric, *A Testimonie of Antiquitie*, which were printed at the order of Parker by John Day in 1566 or 1567. The bishops confirmed with their signatures that the edition was a faithful copy of the original. The *Testimonie* printed Ælfric's Easter Homily (Thorpe, ii. 262–82) and parts

---

[1] Cp. *Catalogues of the Library of Durham Cathedral*, ed. James Raine, London, 1838. The Publications of the Surtees Society, vol. vii. Sisam, *Review of English Studies*, viii (1932), p. 53, says: 'The signature of Leonard Pilkington and a press-mark at the top of the first page indicate that it [MS. G] came to Cambridge from Durham.' But the New Palaeographical Society, which reproduces this press-mark in its First Series, vol. ii, plate 147, 4a, definitely assigns it to Norwich Cathedral, not to Durham. The mark is of cent. xiv–xv.

[2] Printed in 1557 by Robert Stephanus at Paris.

of his letter to Wulfsige and of his second letter to Wulfstan.[1] The first two of these pieces are found in our MS., but since it is defective at the end it may possibly also have contained the third.[2] My guess is that Parker sent this MS. to Pilkington so that he might see for himself before he testified to the authenticity of the printed text. Pilkington may have neglected to return the book and later, at Parker's request, have given it to Cambridge (together with nineteen books of his own) in the same year (1574) in which Parker's own donation of forty volumes came in. This surmise explains best, I think, how Pilkington came to possess a book so different in character from the rest of his collection, and so foreign to his interests.

The language of MS. G is West Saxon. Where it was written is not known.

De Temporibus Anni is found on fols. 255r–261v. This is both the oldest and the best MS. of our text and therefore forms the basis of the present edition.

(8) *Vatican Library, MS. Reginense Lat. 1283.* (*H*)

Only the first volume, covering codices 1–250, has so far been issued of Wilmart's catalogue of the Latin MSS. that were given to the Vatican by Queen Christina of Sweden. MS. 1283, a miscellany, contains on fol. 114v two short extracts from *De Temporibus Anni*, viz. chapters iv. 31–3 and i. 19–21. They are preceded by *Sententiae Hieronymi de utilitate grammaticae artis*, and occupy seven lines. They were discovered by Steinmeyer and published by him in 'Angelsächsisches aus Rom', *Zeitschrift für deutsches Altertum*, xxiv. 191 sq. The discovery being after Cockayne's time, he could not use them for his edition.

This is the only MS. which I have not seen and collated for myself. The variant readings from H in the present edition are taken from Steinmeyer.

---

[1] Fehr, *Die Hirtenbriefe Ælfrics*, p. xxviii.
[2] Ibid., p. 24.

§ 2. SYNOPSIS OF MANUSCRIPT TRADITION

Chapters.	Manuscripts.	Additions.	Omissions.
i. 1–39	ABDG	19–21 also H	
ii. 1–5	ABCDG		5 deest C
iii. 1–29	ABCDG		27–29 desunt C
iv. 1–54	ABCDFG	31–33 also H	2 deest D
			39–40 desunt A
v. 1–9	ABCDFG		
vi. 1–26	ABCDFG	1–5 also E	2–6 desunt C
			13 deest C
			16 deest ABC
			17 deest B
			22 deest C
vii. 1–9	ABCDFG	3–9 also E	
viii. 1–15	ABCDFG	1–12 also E	
ix. 1–14	ABCDFG		
x. 1–24	ABCDFG		3 deest A
			24 deest CD
xi. 1–12	ABCDFG		5–12 desunt F
xii–xiv	ABCDG		

The table sets forth the MS. authority for every part of the text. In the first column, the chapters are those noted in the MSS. themselves, and the paragraphs those marked by MS. G and numbered by the present editor. It appears that MS. G alone offers the complete text.

§ 3. THE RELATIONSHIP OF THE MANUSCRIPTS

The two short fragments in H (chapters iv. 31–3 and i. 19–21) do not offer any clue as to their relation to the other MSS. MS. H is therefore disregarded in the following comparison. In the quotations, minor orthographical or phonological differences are suppressed, where possible in favour of the spellings of MS. G, in order to bring out more clearly the main point of difference

1. MS. G has no mistakes in common with the other MSS. It derives, therefore, from the original, independently of them.

2. It cannot, however, itself be the original since it shows some errors and omissions:[1]

(_a_) The heading of chapter i is lacking in G as well as in

[1] It follows that it would be dangerous to accept the language of the

all other MSS. It must have been *De Die*, as is shown on p. lii below.

(b) In the following four instances, G differs from all other MSS. and would seem to deviate from the original:

ii. 2. þær ðæs (ABCD; ðær G) emnihtes circul is geteald.[1]
vi. 16. se *lengsta* (DF ; *deest* G) dæg.[2]
viii. 11. *hine* ABCDEF ; *hit* G.[3]
xi. 9. hit bi∂ *þurh* (ABCD ; *deest* G) þære sunnan hætan . . . awend.[4]

(c) In the following instances, G differs from all other MSS., but no conclusion is possible as to which tradition is nearer to the original:

i. 22. *mid* (ABD ; ðurh G) hire micclan leohte.

Homilies in Thorpe's edition, which is based on G, as the language of Ælfric himself.

[1] Ælfric always uses the nominative *seo emniht* (cp. chapter vi) but the genitive *ðæs emnihtes* (cp. chapters ii, vi. 7 and vii. 5), giving the word its rightful Old English gender in the nominative, but inflecting it as a masculine or neuter noun. The dative and accusative singular mostly show the ending of the strong (ā) feminines: *emnihte*. Cp. vi. 4 and 6 for the dative and iv. 37, 39, 45, 46 for the accusative. *Ymbe ðas emnihte* (ii. 4) may be accusative singular or plural. Byrhtferð, too, uses the strong dative, *æfter (fram) þære emnihte* (ed. Crawford 84, 12 ; 136, 28 ; 144, 10 ; 168, 15). The nominative also occasionally ends in *-e*. Clark Hall's *Dictionary*, 3rd edition, p. 101, gives this form and describes it as neuter, but places a question-mark against it. I know of two examples: in our text MS. F writes *emnyhte* (vi. 2), and in MS. Harley 3271 *emnihte* is found (printed in *Studien zum altenglischen Computus*, p. 72, line 1). It is possible, of course, to assume that an OE. neuter was formed on the analogy of Latin *aequinoctium*, and that it was declined like *rīce* (N.D.A. *emnihte*, G. *emnihtes*). But it is more likely that *emniht* accepted the forms of the strong feminines in the dative and accusative and that the inflected form occasionally penetrated into the nominative. As for the genitive ending in *-es*, it is frequently found with the strong feminines (Sievers, § 252, note 2). In this particular case it may also be influenced by the adverbial form *nihtes* (Sievers, § 284, note 4) which still exists in modern German (*die Nacht*, but *nachts* on the analogy of *tags*). The reading of ABCD is, then, preferable since it is grammatically correct and supported by other passages. The scribe of G must have repeated the word *ðær* by mistake, although it is barely possible that he *meant* to write þær ðær, 'in the place where' (cp. Jost, *Anglia*, li. 102, note 3). But then *emnihtes* would lack an article. Cp. the reading of MS. A in iv. 22, which uses *þætþæt* for the simple relative.

[2] Only these three MSS. contain the passage. Omission in G is probable, as it writes *lengsta* in vi. 15 and vi. 17.

[3] The pronoun stands for *þone monan*.

[4] *þurh* is added in G by a modern (Thorpe's or Cockayne's ?) hand.

i. 27. We magon *þeah* (ABD; *deest* G) hwæðere to-
cnawan.

ii. 1. þurh ðæs . . . emnihtes *dæg* (ABCD; *dæge* G).

iv. 10. Eahteðe *is* (G; *deest* ABCDF) Scorpius.

iv. 24. gif he urne swa up swa seo sunne *deð* (ABCDF;
deest G).[1]

iv. 26. *nyðemest* ABCD, *nyðemæst* F; *nyðemyst* G.

iv. 34. On swa hwilcum . . . monðe swa *swa* (G; *deest*
ABCDF) se mona geendað.

iv. 49. se langigenda *dæg* (ABCDF; *deest* G).

vi. 4. be ðam *soðan* (ABDEF; *soðum* G) regole.

vi. 7. *langnysse* ABCDF; *langsumnysse* G.

vi. 19. *hatte* ABCDF; *atte* G.[2]

vi. 21. under *dæg* and *niht* ABCDF; under *dæge* and
nihte G.

vii. 8. *ungeteald* AC; *unteald* BDEF; *unateald* G.

viii. 10. *ontent* ABCDF; *atent* G.

ix. 9. *standað* ABCDF; *standan* G.

xii. 1. swa *hattra* (ABCD; *hattre* G) sumor.[3]

The subjunctive is used in MS. G in vii. 2, ix. 9 and xi. 3,
whereas ABCDF (C omits vii. 2) substitute the indicative.
The subjunctive forms must be considered the author's own,
as is shown by i. 32, iii. 6, ix. 14, and x. 24, where several (in
x. 24 *all*) MSS. retain the subjunctive. G employs much more
frequently than the other MSS., but by no means always, the
weakened form of the pronoun, *ðan* instead of *ðam*, especially
in the combinations *ær ðan ðe* and *for ðan ðe*, but also for the
article *ðan*; e.g. *to ðam* (ABCDEF; *ðan* G) *æfene* (viii. 4);
of þam (ABCDF; *ðan* G) *earde* (vii. 1).

3. The remaining six MSS. must all belong to the same
branch of MS. tradition even though they do not share a
single definite mistake. This is proved by the following considerations. ABF have a bad scribal error in x. 1: *Þeos lyft*
. . . *is an ðæra feower* (G; *feorða* ABF) *gesceafta*. D omits

---

[1] The use of the pro-verb would seem to be Ælfric's habit; cp. iii. 10.

[2] Cp. Sievers, § 217, note 1.

[3] Cp., however, ibid.: swa *mara* (AB; *mare* CDG) ðunor.

the crucial word. It may be assumed that the very careful scribe of D found *feorða* in his original and that he omitted the word because it did not make sense. He thus restored a meaning to the passage, if not quite the author's meaning. MS. C lacks nine words in this place. The scribe jumped a line, the words *feower gesceafta* occurring both in x. 1 and x. 2. However, C is closely linked with A (see no. 7 below) and must, therefore, derive from the same original. MS. E stops at viii. 12. It is related to D (see no. 5 below) and thus must also belong to the group *ABCDEF.

4. Within this group, MS. B stands by itself. In some important points B agrees with G, and since G is the only MS. whose tradition is quite independent of that of all others, agreement of G with any other MS. must always be taken as proof for the genuineness of their reading as against that of the other MSS.

Most important of all, BG retain chapter i at the beginning, whereas it is placed at the end in AD. The first chapter is omitted in MS. C, which makes it probable that C, too, derives from a text that placed chapter i at the end. The copyist mistook it for a different work because of the rubric and its emphatic beginning, *Ic wolde eac*. F lacks the first three chapters. MS. E contains merely parts of chapters vi and vii, but since it is related to D (see below no. 5) its prototype may be assumed to have belonged to the tradition which placed the first chapter after the last. The source of H may also have belonged to this group, for in H a short extract from chapter iv. precedes the one from chapter i.

In MS. G alone the text of *De Temporibus Anni* is preceded by a short prologue (or rather it is an epilogue to the Homilies which end in this place) entitled *Oratio*. MS. B retains the last sentence of this text but places it after the first paragraph of chapter i, i.e. after the word *middaneardes*.

The independence of B is further shown in the following places:

i. 38. *cumendne* BG; *cumende* AD.

iii. 25. *ærmergen* BG; *ærne mergen* ACD.

iv. 21. seo sunne and *se* (ACDF; *deest* BG) mona.[1]

xi. 2. *blædum* BG; *blæde* AC; *blæd* D; *blæda* F.

The foregoing four quotations constitute a group *ACDF.
To them must be added E because of its relationship with
D, which shall be proved next.

5. DE form a closely related group differing from ACF.
Considering the shortness of the two fragments E, the points
of agreement are all the more striking.

vii. 3. *and* (ABCFG; *deest* DE) eft on merigen.

vii. 7. setton . . . *to* (ABCFG; *on* DE) ðam monðe.

viii. 5. Be ðisum is *oft* (DEG; *deest* ABCF) micel embspræc.[2]

Note 1. DE agree (accidentally?) with A:

vii. 4. þam . . . fifand*sixtig* (BCFG; *sixtigum* ADE) dagum.

vii. 5. on . . . fifand*sixtig* (BCFG; *sixtigum* DE) dagum.[3]

Note 2. DE agree (accidentally?) with B:

viii. 3. *setlunge* ACFG; *setlgange* BDE.

It is not improbable that B on the one hand, and DE on
the other, made the change independently, for F replaces
in a similar fashion x. 17 *upspringe* ABCDG by *upgange*. All
MSS. use *setlung* in iii. 18, but *niðergang* in iv. 15 and *upgang*
in i. 24 (twice), iii. 18, 25 and iv. 15.

It should be further noted that BDG alone preserve correctly the name of *Meroe* in vi. 14. MS. A has *Meloe*, C has
Mede and F, *Merode*. E does not offer the passage.

6. D cannot have been the prototype of E:

viii. 5. *gelæredan* ABEFG; *læredan* CD.

viii. 7. *nebið* ABCEFG; *nis* D.

viii. 10. *awend* ACEFG; *onwend* B; *gewend* D.

D cannot derive from E, because the latter consists merely
of two short fragments, and because of the following deviations and errors of E:

vi. 1. on *sancta* (AE; *deest* BCFG) marian mæssedæg.[4]

---

[1] Even though BG would here seem to be in the wrong, the agreement is
probably accidental and can hardly suffice to suggest a group *BG.

[2] This quotation is not compelling proof as it implies that B lost the word
oft independently of ACF.

[3] Lacuna in A.   [4] A alone adds the word *sancta* in vi. 3.

vi. 5. *secgað* ABDFG; *scewiað* E.

vii. 4. *of* (BCDFG; *on* AE) þam six tidum

vii. 5. *Seo* (BCDFG; *se* E) sunne.[1]

vii. 6. tida, an ABCDFG; tida, *þæt is* an E.

vii. 8. *spræc* BCDFG; *sprecð* AE.

Ibid. gerynu *and* (ABCDFG; *deest* E) gif.

viii. title BDFG; *deest* ACE.

viii. 9. *ðwyrs* G; *þwyres* ABCDF; *þwyre* E.

Ibid. gif *heo* (ABCDFG; *he* E) hine ontent neoðan.

Ibid. to ðære *sunnan* (ABCDFG; *sunna* E) weard.

Note. In three of the quotations above E agrees with A, in one it agrees with A and C. The agreements of A and E may be significant.

7. The relation of the remaining three MSS., A, C, and F, is problematical. Possible proof for a common ancestor *ACF might be found in the passage viii. 5, where ABCF omit the word *oft* against the evidence of DEG. However, since MS. B, which stands apart, shows the same omission, the agreement is probably an accident. A close connexion between MSS. A and C is attested by a lengthy list of common errors, changes, and omissions. AF agree in seven places, but in each one of them mere coincidence is likely. The most natural assumption, therefore, would be that F derives from *ACDEF without any closer relation to other MSS. in this group.

(*a*) AC agree in the following points:

The Latin chapter-headings are omitted throughout in AC, excepting that A has the title of chapter ii which, in it, stands first.[2]

iii. 16. *niwum* BDG; *niwne* A, *niwe* C.

iii. 17. *ormæte* BDG; *ormætlic* A, *ormætlice* C.

iv. 13. Endlyfte *is* (BDFG; *deest* AC) Aquarius.

iv. 21 and iv. 30. *tunglan* BDFG; *tungla* AC.[3]

---

[1] Lacuna in A.

[2] E omits the headings of chapters vi and viii, the only chapters which it begins with the first paragraph.

[3] But cp. ix. 6 *tunglan* BCDFG; *tungla* A.

iv. 24. underyrnan ealle ða twelf *tunglan* (BDFG; *tungla* A); eallam þa twel tungla C.[1]

iv. 25. heo is *swiðe* (BDFG; *swa* AC) upp.

iv. 32. þæt gear *þe* (BDFGH; *deest* AC) we hatað Communis *þæt* (C, *þæt þæt* A; *deest* BDFGH) hæfð twelf niwe monan, and þæt gear *ðe* (*deest* AC) *we* (*deest* C) *hatað* (*deest* C) embolismus *þæt* (AC; *deest* BDFGH) hæfð þreottyne niwe monan.

iv. 50. forði *he* (AC; *deest* BDFG) hæfð.

iv. 51. forði *he* (AC; *deest* BDFG) bið.

v. 5. Thirteen words omitted in AC as against complete text in BDFG.

v. 6. *nebihð* BDFG; *nebið* C.[2]

vi. 1. and *þa* (AC; *deest* BDEFG) egyptiscan.

vi. 12. on *nane* (BDFG; *nanre* A, *anre* C) healfe.

vi. 15. AC erroneously anticipate the first five words of vi. 18. BDFG preserve the true reading.

vi. 16. The whole paragraph omitted AC.[3]

vi. 18. on ðam *ylcan* (BDFG; *deest* AC) earde.

vi. 24. An ðæra *dæla* (BDFG; *deest* AC) is.

vi. 26. *sunstedum* BDFG; *sunstede* AC.

vii. 2. stod *stille* (AC; *deest* BDFG) anes dæges lencge.

viii. 6. *ærne* BDEFG; *deest* AC.

viii. 7. be ðam monan wiglian BDEFG; wiglian be þam monan AC.

viii. 12. *cepan* BDEFG; *kepan* AC.

viii. 13. *weaxendum* BDEFG; *weaxenda* A, *weaxende* C.

viii. 14. *treowu* DG, *treowa* BF; *treow* AC.

x. 13. *fornean* BDFG; *forneah* AC.

x. 19. on ledenum *bocum* (AC; *deest* BDFG).

xi. 4. *eorðlice wæstmas* BDFG; *eorðmæstmas* A, *eorð-wæstmas* C.

---

[1] For *underyrnan* with the dative cp. iii. 15, where *underscytan* is used with that case.

[2] Probably confusion with the substantive verb.

[3] B omits this and the following paragraph.

xiv. 1. *hætan* BDG; *wætan* AC.

xiv. 2. bradnysse *and* (BDG; *deest* AC) frecenful.

Note 1. D agrees with AC in one place:

v. 7. þeah *þe* (ACD; *deest* BFG) heo.

Note 2. B agrees with AC in two places:

x. 18. *astyrað* DFG; *astyreð* ABC.[1]

ABC omit vi. 16 which is found in DFG. However, B alone also omits vi. 17, AC are quite corrupt in vi. 15, D gives a wrong numeral in vi. 15, C in vi. 17. The section vi. 15–17 presents some difficulty to the copyist because of its parallels and repetitive phrasing. This makes it probable that B on the one hand, and AC on the other, omitted vi. 16 independently.

(*b*) AF agree in the following points:

iv. 33. *þritig* BCDGH; *þriti* AF.[2]

iv. 44. *winterlicum* BCDG; *winterlican* AF.

vii. 2. seo sunne *ða* (BCDG; *deest* AF) stod.

vii. 6. *An* dæg and *an* niht BCDEG; *on* dæg and *on* niht AF.

viii. 9. he *and* (AF; *deest* BCDEG) gif.

x. 2. þæt *sind* BCDG; þæt *is* AF.

x. 15. *Se wind* BCDG; *desunt* AF.

Note 1. B agrees with AF in one place:

iv. 51. *he ofergæð* CDG; *heo ofergæð* BF; *heo forgæþ* A.

8. Even a casual glance at the apparatus criticus furnishes compelling proof that A cannot have been the source of C. MS. A has a considerable number of errors not found in C and is in every way the least reliable MS.

The reverse assumption, viz. that A derives from C, is prohibited by the late date of C as well as by the fact that the scribe of C (or of its prototype) submitted the text to a more or less systematic revision, omitting a number of paragraphs. There are also errors in C not shared by A.

[1] DFG conjugate according to class II, whereas ABC use the verb correctly as a *ja*-stem of the first class. Cp. Sievers, § 410, 5.

[2] Cp., however, ibid.: *twentig* BCDFGH; *twenti* A.

The descent of the seven MSS. under consideration may be visualized by the following graph:

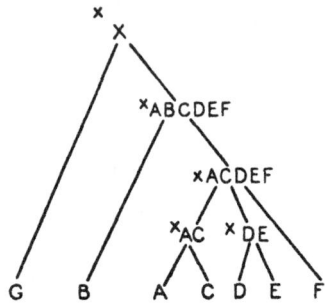

The reconstruction of the relationship of the MSS. now given is in opposition to the following agreements (in addition compare those mentioned in the notes to nos. 5, 6, and 7):

9. CD have a few marked agreements:

iii. 20. Uesperum þonne ABG; Uesperum *þæt is* þonne C, Uesperum *þæt is æfen* þonne D.

iii. 25. þæt is *se* (ABG; *deest* CD) ærmergen.

iv. 29. *forne gean* BFG; *forne agean* A; *forn ongean* CD.

viii. 5. *gelæredan* ABG, *gelaredan* E, *gelæredon* F; *læredan* CD.

x. 24. the whole paragraph omitted CD, but found ABFG.

None of these quotations offers compelling proof for a group *CD.

10. AD share one omission:

iii. 7. ða twelf *tunglena* (BCG; *deest* AD) tacna.

The omission proves nothing, as in iv. 26 MS. A omits the word *tungla* which is found in BCDFG.

11. AG differ from the other MSS.:

x. 6. ðurh ða *lyft* (BCDF; *lyfte* AG).

Cp. Sievers, § 269, note 1, and vi. 4 ða *halgan eastertide* (all MSS.).

12. DF share one omission:

ix. 8. *an* (ABCG; *deest* DF) on suðdæle.

13. BC agree in one spelling:

iii. 5. manncynne ADG; mannkynne BC.

Although this spelling is rare in our MSS., it is not uncommon in West Saxon. Cp. Sievers, § 207, note 2, and viii. 12 *cepan* BDEFG; *kepan* AC.

14. AB agree in a rare spelling and in a rather striking error:

i. 3. *god* DG; *good* AB.

iii. 29. on *sumum* (DG; *sunnan* A, *sunnon* B) dæge þære wucan.

MS. A alone has the same mistake in iv. 24 and changes, in the opposite direction, *sunstede* into *sumstede* in iv. 20 and iv. 38. MS. B offers in the last two places the spelling *sunnstede*. A partial agreement between A and B is found in:

v. 6. heo næfre nebihð *napor* (A; *deest* BCDFG) ne (AB; *deest* CDFG) ufor ne neoðor.

The general conclusion must be that the quotations given in nos. 9–14 are of little force and cannot disprove the relationship of the MSS. as established in nos. 1–8.

Our investigation of the MSS. further shows that the MSS. now extant derive from lines of descent which must have separated quite early. All of them have a number of definite, sometimes pretty bad scribal errors, but they have very few mistakes in common. (The only exception is AC which, for this reason, is the only group that can be proved beyond any doubt.) In its turn, this shows that the number of MSS. must have been considerable in the eleventh century. If this were not so, the MSS. now preserved could not go back to so widely divergent branches of tradition but would show closer interrelation or actual dependence upon one another.

§ 4. EARLIER EDITIONS AND TRANSLATIONS

About the middle of the last century three complete editions of *De Temporibus Anni*, as well as a selection, were published within the space of twenty-five years. Since then nothing has appeared except a German translation of the last five chapters. It is not necessary, therefore, to offer an apology for the present attempt to provide an edition based on modern principles of textual criticism. The editions referred to are the following:

(1) Thomas Wright, *Popular Treatises on Science Written During the Middle Ages*, London, 1841, pp. 1-19.

This, the *editio princeps*, is based on MS. B. The editor also knew MSS. ADFG but did not print any collations. No attempt is made to correct the errors of B, excepting that the two sentences vi. 16-17 which are lacking in B are supplied from MS. D. An English translation is added.

(2) Louis F. Klipstein, *Analecta Anglo-Saxonica*, New York, 1849, vol. i, pp. 245-7.

Offers five extracts, viz. iii. 3-4; iii. 18-25; iv. 27-8; v. 1-9, and vi. 9. The text is not copied from Wright's edition but must have been taken from MS. D.

(3) C. W. Bouterwek, *Screadunga. Anglo-Saxonica maximam partem inedita*, Elberfeld, 1858, pp. 23-31.

The text is based on MS. G, as is the present edition. In his preface the editor states that Benjamin Thorpe communicated to him the collations of G with Wright's edition, and that the variant readings of B (which he prints at the foot of his pages) were collated afresh with the MS. The indirect way in which he obtained his text may account for Bouterwek's many inaccuracies. It is noteworthy that the text not infrequently offers the readings of B instead of those of G, a fact which is also explained by the manner in which it was composed.

(4) Oswald Cockayne, *Leechdoms, Wortcunning, and Starcraft of Early England*, vol. iii, London, 1866 (Rolls Series), pp. 231-81.

This has been the accepted edition for more than seventy years. Cockayne does not say which MS. he followed, nor does he state what principles of selection guided him in preparing his edition. It is, however, based on Wright's edition of B,[1] but offers variant readings from ADFG as well. Neither text nor variant readings are very accurate, and the latter are quite incomplete. In a number of places the reading of other MSS. is given in the text and that of B relegated to the *apparatus criticus*. This is done sometimes to expunge an error of B, sometimes merely to please the editor's fancy.[2] There is no system in these emendations, as indeed there could not be, since Cockayne made no attempt to establish the relationship of the MSS. or to find out which was the best. He failed to correct all mistakes of B but, on the other hand, did not adhere resolutely to this MS. either. The readings of B are retained very often against the evidence of all other MSS., and sometimes a reading is adopted which is not supported by any MS. authority whatsoever, as iii. 28 *lunaris* for *luna* and vi. 5 *ræderas* for *fæderas*. Yet Cockayne's edition marks a considerable progress when compared with Wright's, who printed B with all its errors and shortcomings and even misread or misrepresented the MS. repeatedly.

Cockayne's translation is also superior to Thomas Wright's. His style, it is true, is somewhat stilted and suffers from a number of far-fetched archaisms, but he corrects most of Wright's mistranslations, whether they are due to the latter's faulty text or to his inadequate knowledge of Old English. It was, therefore, thought unnecessary to add a translation to the present edition. Where Cockayne's translation goes astray it is usually due to some fault in his text (which will become obvious when compared with ours) or to his inclination to translate too literally. I note a few examples:

i. 1. 'courses of the year' should be 'course of the year'.

i. 7. 'beyond the discussion . . . of men' should be 'invisible'.

---

[1] Of the many proofs of this I will quote only the fact that on p. 237 Cockayne italicizes (to show that he is filling a gap) the words *they seem*, translating ðincð, which is found in all other MSS. and in his own text, but is missing in B and in Wright's edition.

[2] As, for instance, when he prefers in vi. 15 the obviously corrupt reading of A against the evidence of BDFG.

i. 11. 'in equilibrium' should be 'equally'.

i. 27. 'know' should be 'recognize'.

iii. 7. 'between us and the broad circle' should be 'because of the breadth of the circle'.

iv. 41 and vi. 26. 'this sphere' should be 'our globe (the earth)'.

vi. 2. 'ascertaineth for us about' should be 'instructs us about'.

vi. 5. 'the faithful advisers' should be 'the Fathers of the Church'.

Ibid. 'sure day measurements' should be 'reliable sun-dials'.

viii. 5. 'distinction' should be 'rule'.[1]

viii. 11. 'explanation' should be 'rule'.[2]

ix. 5. 'incredible' should be 'unchristian, heretical'.

(5) Gustav Hellmann, *Denkmäler mittelalterlicher Meteorologie*, Berlin, 1904, pp. 20–2.

Contains a German translation of chapters x (partially), xi–xiv, based on Thomas Wright's edition.

§ 5. THE AUTHOR

Two MSS. give the name of an author for our treatise. MS. B has *Incipiunt Pauca De Temporibus Bedae Presbiteri.* This is simply the scribe's paraphrase of the first sentence of the text and cannot be taken as proof that Bede himself prepared the OE. adaptation from his three Latin books on science. MS. F is entitled by a much later hand *Æthelardus de Compoto.* Thomas Wright[3] assumed that Athelard of Bath was meant by this attribution and rejected it because the latter lived in the twelfth century. Wanley thought there might have been more justice in ascribing the contents of fols. 142–3 of Caligula A XV[4] to this author. There is no need, however, to be too contemptuous of Wan-

---

[1] Cp. *Anglia*, lviii. 300.   [2] Ibid.

[3] *Popular Treatises on Science*, London, 1841, p. viii.

[4] See above, p. xxiii.

ley's *quidam Neotericus* who made the attribution. He may have meant the historian Æthelweard, great-grandson of Æthelred, the brother of Alfred.[1] This Æthelweard was ealdorman of the western provinces and the author of a Latin chronicle extending to the year 975.[2] The chronicle contains a text of the Ages of the World which is also found in this same MS. Caligula A. XV and in Ælfric's Homilies.[3] He died in 998 or in 1001.[4] To Æthelweard are addressed Ælfric's prefaces to his *Lives of Saints*[5] and his *Preface to Genesis*,[6] whilst his son Æthelmær is mentioned in the former as well as in the Old English preface of the Homilies.[7] The father and especially the son were Ælfric's friends and benefactors.

There is no evidence to support the attribution of *De Temporibus Anni* to Æthelweard, but if the present interpretation is allowed the *Neotericus* did not go as far astray as was supposed by Wright.

Ælfric's authorship was conjectured quite early; Wanley never doubted it, Dietrich[8] supported it with some fresh argument, and Reum[9] devoted a special monograph to the

[1] Cp. *Encyclopaedia Britannica*, 13th edition, i. 291, and *Dictionary of National Biography*, xviii. 35 sq.

[2] Dietrich, *Abt Aelfric*, p. 222, leaves the question undecided if Æthelweard dux and the author of the chronicle were the same person, whereas C. L. White, *Ælfric, A new Study of his Life and Writings*, Boston, New York, and London, 1898, p. 48, note 3, is 'nearly certain' that the identification is justified. W. G. Searle, *Anglo-Saxon Bishops, Kings and Nobles*, Cambridge, 1899, p. 435, calls Æthelweard dux 'the Chronicler', and the Introduction to *The Exeter Book*, London, 1933, p. 88, says that 'Æthelweard has generally been accepted as identical with the historian of that name'.

[3] The three texts do not, however, belong to the same tradition. See M. Förster, 'Die Weltzeitalter bei den Angelsachsen', *Die Neueren Sprachen*, 6. Beiheft, Marburg, 1925, pp. 195, 192, and 187.

[4] According to K. Sisam, op. cit., p. 55, note 3, and O. Cockayne, *Leechdoms*, iii, p. xv, respectively.

[5] Ed. W. W. Skeat, E.E.T.S., O.S. 76 (1881), p. 4.

[6] Ed. S. J. Crawford, E.E.T.S., O.S. 160 (1922), p. 76.

[7] Ed. B. Thorpe, London, 1843, i. 2. Æthelweard, too, is mentioned there; Thorpe, i. 8.

[8] Eduard Franz Dietrich, 'Abt Ælfric. Zur Literaturgeschichte der angelsächsischen Kirche', Niedner's *Zeitschrift für historische Theologie*, 1855, p. 494.

[9] A. Reum, 'De Temporibus. Ein echtes Werk des Abtes Ælfric', *Anglia*, xi. 457–98.

question. The only piece of external evidence, viz. that *De Temporibus Anni* is appended to Ælfric's Homilies in MS. G, was mentioned by Cockayne,[1] who added this pertinent comment: 'the evidence that the work is Ælfric's arises from this circumstance only and a general probability from the method of handling the translation from the Latin, with the difficulty of assigning such a work to any other writer.' Reum fastened his attention on the second and third of Cockayne's arguments. He analysed Ælfric's treatment of his sources and his linguistic peculiarities, and showed that *De Temporibus Anni* exhibits characteristics similar to those of Ælfric's acknowledged writings. Reum's discussion of Ælfric's language is somewhat pedantic and suffers from an insufficient knowledge of the traditions on which the Anglo-Saxon built. When Ælfric compares the three qualities of the sun with the Holy Trinity,[2] or the subterranean streams with the veins of the human body,[3] or the earth with a fir-cone,[4] he is not using the figurative language of the clever pedagogue, as Reum supposes.[5] These comparisons can be found in Ælfric's sources and go back to an ancient tradition of theological or cosmic speculation.[6] How easy it is to go astray in deciding authorship on the strength of stylistic evidence is also shown by Reum. In discussing[7] the passage on the epacts in MS. Caligula A. XV which he takes to be an appendix to *De Temporibus Anni*, he finds a similarity of expression with Ælfric, whereas in actual fact the passage is Byrhtferð's.[8]

All the same, however inconclusive Reum's individual arguments may be, their cumulative weight is, I think, sufficient to establish Ælfric's authorship. This is especially true of the parallels which he adduces between *De Temporibus Anni* and the other works of Ælfric. These parallels, together with

---

[1] Op. cit., p. xix. Wülcker, *Grundriss*, p. 478, is mistaken in saying that the treatise is surrounded by works of Ælfric in all MSS.

[2] *Homilies*, ed. Thorpe, i. 282.   [3] *De Temporibus Anni*, v. 8.

[4] Ibid. vi. 9.   [5] Op. cit., pp. 484 and 486.

[6] Cp. my article in *Anglia*, lviii (1934), pp. 294 and 314.

[7] Op. cit., p. 495.

[8] *Studien zum altenglischen Computus*, pp. 52–4.

many more which Reum overlooked, will be found in the
present edition appended to their appropriate passages.
The only scholar who, in more recent times, denied Ælfric's
authorship of *De Temporibus Anni* was Eduard Sievers.[1]
According to him, *De Temporibus* represents a vocal type
differing from Ælfric's, while the *Hexameron* is different again
and should be attributed to a third author. The *Heptateuch*
he described as an older translation revised by Ælfric. It is
impossible to take issue with Sievers whose method of *Schall-
analyse* must be either accepted or rejected *in toto*. Stylistic
evidence will not avail against him, as he would probably have
answered that a style can be learned or copied, but that a
voice cannot.

De Temporibus yields a piece of biographical information
which, as far as I am aware, has not been utilized by any
of Ælfric's biographers. Yet it may be of importance. In
chapter vi. 18, when speaking of the brightness of summer
nights in northern England, Ælfric adds the remark (not
found in his source) that 'we ourselves have very often seen
this'.[2] If we take the passage at its face value (as indeed we
must, since the brightness of midsummer nights is certainly
much more remarkable in the north of England than in the
south) it establishes the fact that Ælfric spent some time of
his life in northern England. According to Dietrich, Ælfric
was born about 955. *De Temporibus Anni* was probably
written in 993, so that Ælfric's sojourn in the north would fall
between these two dates, or in the earlier part of his life. We
are fairly well informed about his whereabouts after he had
entered Æthelwold's monastic school at Winchester, which
event Dietrich places in about 971. Unless we assume that
Ælfric went north on a journey or some mission similar to the
one which took him to Cernel (which seems unlikely because
he would have mentioned it in his Prefaces), the only conclusion can be that his home was in the north of England,

[1] 'Metrische Studien IV', *Abhandlungen der Sächsischen Gesellschaft der
Wissenschaften*, xxxv, Leipzig, 1918, pp. 211–19. Cp. especially p. 212,
note 1.　　　[2] See p. 94 sq., below.

and that he was brought up there before being sent south for his education. Only two facts can be gathered from his writings regarding his life before he became the pupil of Æthelwold: He repeatedly praises the time of Edgar (King of the Anglians since 957, of all England since 959, died in 975) as the happiest he has known;[1] and in the Preface to Genesis he speaks of an ignorant priest who was his teacher. The best we can say is that these pieces of information do not conflict with the assumption that Ælfric came from the north. The evidence of the one remark in *De Temporibus Anni* is, of course, far too slender to base definite conclusions on it. But if Ælfric was a north-country boy, the fact may be betrayed by occasional lapses into Anglian, even though he certainly was taught to write West-Saxon at Winchester. A study might well be made of his writings, especially the earlier ones, with a view to ascertaining if there are Anglian peculiarities of speech in MSS. which are demonstrably of southern origin. If so, they would be Ælfric's, not the scribe's. Books such as R. Jordan's *Eigentümlichkeiten des anglischen Wortschatzes*, Heidelberg 1906, are of no assistance in deciding this question since they are based on the assumption that what is Ælfric's is *ipso facto* West-Saxon.

§ 6. THE DATE

Wülcker and Skeat[2] classed our tract with the writings of Ælfric that cannot be dated. There are, however, some facts which indicate at least the approximate date.

(1) Byrhtferð's *Manual*, written in 1011,[3] contains numerous quotations from *De Temporibus Anni*. The latter must therefore be older.

(2) In his version of the *Interrogationes Sigewulfi* Ælfric

---

[1] Dietrich, Part II, p. 243.

[2] R. Wülcker, *Grundriss*, pp. 476 sqq.; Skeat, Preface to volume ii of his *Ælfric's Lives of Saints*, p. xxxvii.

[3] Cp. Cockayne, *Leechdoms*, iii, p. xxx; K. M. Classen, op. cit., p. 19.

refs to *De Temporibus Anni* as an earlier work.[1] Unfortunately, the date of the *Interrogationes* is very uncertain.

(3) By general agreement the *Homilies* are Ælfric's first product. They are dated about 990–1 for the First Series and about 992 for the Second Series. The former date has been accepted since Dietrich's investigation, the latter was established (instead of the earlier assumption, 994) by K. Sisam.[2] Reum[3] thought he could prove that *De Temporibus Anni* was composed after the First Series and before the Second, i.e. A.D. 991. His argument, however, rests on a misunderstanding[4] of a passage in Cockayne's Preface which led him to believe that MS. G inserted our treatise in between the First and Second Series of Homilies. In actual fact it is found there *after* the Second Series. It is true, though, that *De Temporibus Anni* has more numerous correspondences with the Homilies, especially the homily *Octabas et Circumcisio Domini Nostri* in the First Series, than with any other of Ælfric's works. Both because of its position in MS. G (following the Homilies and preceding Ælfric's Letter to Bishop Wulfsige) and because of certain parallels in the two texts, it seems probable that *De Temporibus Anni* was written immediately following the Second Series of the Homilies.

(4) There is further evidence in favour of this assumption. MS. D of *De Temporibus Anni* is the only one of which we know definitely that it comes from Winchester. It was written there almost certainly after Ælfric had left for Eynsham (1005), perhaps even after his death (between 1020 and 1025). Nevertheless, the tradition and influence of his works must have been strong in that centre. If, therefore, we find a short colophon added to D which agrees very closely with the last sentence of the Homilies (Second Series),[5] it is fair to assume that it was not invented by the scribe, but taken from the earlier work. In that case, the exemplar of D contained

---

[1] Cp. MacLean, *Anglia*, vi. 468; Henel, *Anglia*, lviii. 297 sqq.
[2] Op. cit., p. 67 sq.   [3] Op. cit., pp. 496–8.   [4] Ibid., p. 459.
[5] See above, p. xxi. The end of the *Interrogationes* may also be compared, but it is not worded quite in the same fashion.

both the Homilies and *De Temporibus Anni*, and treated the latter as an integral part of the Homilies by transferring their last sentence, the benediction, to its end.

It looks as if Ælfric had meant to guard against just such a contingency when he prefaced *De Temporibus Anni* by the remark, 'Hereafter follows a little piece on the Times of the Year, which is not accounted a sermon, but is rather to be read by anybody whom it pleases.'[1] This sentence is found in MS. G, where our tract is preceded by the Homilies. It is also found partially, and placed after the first sentence of the text, in MS. B, which does not contain the Homilies. Now B and G are the only MSS. contemporary with Ælfric. The inference is that both go back to a book which contained the Homilies as well as *De Temporibus Anni*, that this book was written under Ælfric's supervision, and that it had his warning against confusing the Homilies, which are sermons, with *De Temporibus Anni*, which is not a sermon.

Besides BG, A and D are the only other MSS. that offer all chapters of *De Temporibus Anni*. In these, however, the first chapter is placed at the end. This was done, no doubt, because the end of this chapter with its mystical interpretation of sun, moon, and stars made a finer peroration for a sermon than does the sober exposition of meteorological phenomena with which the treatise ends in the original (i.e. Ælfric's) arrangement. The re-grouping cannot be the work of Ælfric himself, because it is carelessly done. Both MSS. fail to delete the colophon, 'Let this narrative be thus here ended', at the end of chapter xiv. D alone attempts to smooth out the transition to the next chapter (really the first) by omitting the first fourteen words with the tell-tale phrase, 'I would *also* . . .'. But what remains of the sentence in D is palpably incomplete, lacking an object.

To sum up: The four complete MSS. of *De Temporibus Anni* all go back to copies wherein the Homilies were found as well as our treatise; with this difference: B and G, coeval with the

---

[1] Mr. Sisam's translation.

author, derive from a tradition which kept the two works separate, at his express direction. A and D, written after Ælfric's time, show that their prototype regarded *De Temporibus Anni* as the last sermon in the Second Series of Homilies. D has more definite traces of this contamination than A.

(5) There is a parallel to this process in the incorporation of the *Interrogationes* into the *Lives of Saints*. It is the best MS. of the *Lives*, Cotton Julius B. VII, which admits the *Interrogationes* into the Third Series of sermons. MacLean[1] aptly compared the *De Temporibus Anni* with the *Interrogationes* by calling the one the 'Popular Treatise on Science',[2] and the other the 'Popular Treatise on Theology'. Both were written by Ælfric to elucidate further certain matters touched upon in his collections of sermons. The former was meant as an appendix to the first two series (the *Homilies*), the latter to the Third Series (the *Lives*).

The *Lives* were written some time between 996–8, or, according to Sisam,[3] between 993 and 998. Since the Second Series was completed in 992, the date of *De Temporibus Anni* is 992 at the earliest, and 998 at the latest. Time must be allowed for the production of so lengthy a book as is the *Lives*, and thus A.D. 993 would seem a probable date.

(6) A serious objection may be made to the foregoing argument. The *Grammar* has always been taken to be Ælfric's second work, since in its preface reference is made to the two books of Homilies, but to no other work. Now in the Latin preface to his *Lives* Ælfric states: *Non mihi imputetur quod diuinam scripturam nostrae lingue infero . . . ; sed decreui modo quiescere post quartum librum a tali studio, ne superfluus iudicer.*[4] Dietrich, Wülcker, Sisam, indeed all writers on the subject that I know of, interpret this to mean that the *Lives*

---

[1] *Anglia*, vi. 427. [2] Following Thomas Wright.
[3] Op. cit., p. 55, note 3: 'I incline to place the *Lives* early in the period 993–998.' Sisam does not mention, and consequently fails to refute, the argument that the *Lives* are not dedicated to Archbishop Sigeric, as were the *Homilies*, and that for this reason their completion should fall after A.D. 995. [4] Ed. W. W. Skeat, p. 4.

are the fourth book from Ælfric's pen. As he always considered his *Homilies* as two books, the *Grammar* would be the third, and there would be no room for *De Temporibus Anni* before the *Lives*.[1] But it is possible to take a different view. In the first place, when Ælfric referred to the *Lives* as his fourth work, he need not have counted *De Temporibus Anni* as a separate book. He may have thought of his major works only, disregarding this short treatise appended to the *Homilies*. Secondly, the context makes it much more natural to assume that Ælfric referred, not to his fourth book generally, but to the fourth book of translations from sacred writings. He has scruples about these translations and says that he will desist from *such* pursuits after the completion of the fourth book. Nothing is said about writings on secular or learned matters, and nothing indicates that Ælfric intended in this passage to give a chronology of his works. Now if the *Lives* are the 'fourth book' of translations from sacred sources, the difficulty arises that we do not know of a 'third book' of such content, unless it be the translation from the Old Testament, the OE. *Heptateuch*. The authorship of the *Heptateuch* is in doubt. Dietrich and Sievers[2] held that most of it was the work of an earlier translator, revised by Ælfric. K. Jost,[3] however, maintains as the result of a most painstaking study that only *Genesis* i–xxiv and *Numeri* xiii–xxvi are now preserved in the form which Ælfric gave them. The rest of the OE. *Pentateuch* he considers the work of a later redactor, based upon translations or biblical homilies by Ælfric. Dietrich gave to the *Heptateuch* fifth place (after the *Lives*) in the order of Ælfric's writings and dated it A.D. 997. Neither Jost nor Crawford in his edition[4] raises the question of the date of Ælfric's biblical translations, and there is no compelling proof that they followed the *Lives*.

---

[1] Miss White raised this difficulty p. 54, note 1.   [2] Cp. p. xlv, above.
[3] Karl Jost, 'Unechte Ælfrictexte', *Anglia*, li. 81–103; 177–219.
[4] E.E.T.S., O.S. 160 (1922).

However this may be, enough has been said, I think, to show that there is room for assigning minor works, especially works of non-sacred content, to the period between the *Homilies* and the *Lives* without conflicting with the available evidence. And thus the date of about A.D. 993 for *De Temporibus Anni* may stand.

(7) There remains only the question of the relative position of the *Grammar* and of *De Temporibus Anni*. Mr. Sisam, if I read him aright, at first repeats the traditional view that the *Grammar* is Ælfric's second work. But on the very next page,[1] on the evidence of his own investigation of certain matters of syntax, he groups *De Temporibus Anni* with the *Homilies*, and the *Grammar* with Ælfric's later works. To challenge Dietrich's views, and in order to arouse further discussion of the subject, I will tentatively arrange the earlier works of Ælfric as follows:

i, ii. The two Series of *Homilies*.

ii *a*. An appendix[2] to the Second Series: *De Temporibus Anni*.

iii. The *Grammar*.

iv. Portions of the Old Testament: *Genesis* i–xxiv and *Numeri* xiii–xxvi; perhaps more.

v. The *Lives of Saints*.

v *a*. An appendix to the same: the *Interrogationes*.

The Prefaces to the *Homilies*, *Genesis*, and the *Lives* mention the name of Æthelweard dux. These works all belong to the first, and most productive, period of Ælfric's authorship. When it was ended, he finally adhered to his determination not to translate any more sacred writings, an intention which he had expressed as early as in the Prayer which follows the Second Series of Homilies in MS. G.

---

[1] Op. cit., p. 66 sq.

[2] That *De Temporibus Anni* really was meant as an appendix, and not as a separate work, is also attested by the fact that Ælfric failed to give his name in the prefatory paragraph. *Ic wolde eac . . . gadrian*, he says, not: *Ic Ælfric*. In most of his major works he puts his name. Cp. C. L. White, *Ælfric*, Boston, 1898, pp. 165–82.

§ 7. THE TITLE

Ælfric's short treatise on natural science has come down to us under the name of *De Temporibus Anni*. This is the title that two MSS. (D and G) show at the head of the first chapter,[1] but since it does not fit this chapter it must be assumed that it was really meant for the book as a whole, and that it replaced the original chapter-heading.[2] The title was taken over from one of Ælfric's sources, Bede's Latin treatise *De Temporibus*. The latter's more comprehensive treatment of related subjects, used even more extensively by Ælfric, commonly bears the name of *De Temporum Ratione*. But it, too, was known as *De Temporibus*, as is witnessed by Byrhtferð, who mentions[3] *librum reuerentissimi Bedę quem De Temporibus pretitulauit* when he refers clearly to the larger work. Ælfric himself, in the opening sentence of his work, describes it as a gathering of information, i.e. an extract, from Bede's book *be ðæs geares ymbrenum*.[4] The Old English phrase, then, was Ælfric's name for his book, it was almost identical with the name that was given to one, and occasionally both, of Bede's books on science, and it was, therefore, rendered accurately enough by the scribes when they gave the title as *De Temporibus Anni*.

The heading of the first chapter, which in the MSS. had to yield its place to that of the whole book, must have been *De Die*. This is proved by its contents as well as by the fact that the succeeding chapters are called *De Primo Die Saeculi*, *De Nocte*, *De Anno*, &c. Ælfric's chapter-headings are taken

---

[1] B has *Incipiunt Pauca de Temporibus Bedae Presbiteri*. MS. A has no title. CEF do not offer the first chapter.

[2] In a similar fashion MS. A, which omits all other chapter-headings, retains the one for the second chapter which, in it, is the opening chapter. Thus the scribe of A (or of its original) must have mistaken the title of chapter ii for that of the whole text.

[3] Ed. Crawford 226, 13. Cp. ibid. 74, 19.

[4] The short preamble in MS. G calls it *án lytel cwyde be géarlicum tidum*. Cp. also Ælfric's *Homilies*, ed. Thorpe, i. 98: *Þa ealdan Romani . . . ongunnon þæs geares ymbryne on ðysum dæge* (i.e. January 1st).

from Bede's.[1] In the latter's *De Temporibus* chapter ii
is entitled *De Die*; iii, *De Nocte*; ix, *De Anno*. In his *De
Temporum Ratione* the sequence is as follows: chapter v,
De die; vi, *Ubi primus dies seculi sit*; vii, *De nocte*.

§ 8. THE SOURCES

In the first paragraph of his treatise Ælfric acknowledges
his indebtedness to Bede, and he mentions him again in ii. 4
as his authority for the right date of the vernal equinox. The
only other writer to whom he refers by name is Augustine
(vii. 8), but here he merely follows his source, Bede's *De
Temporum Ratione*, cap. xxxix.[2] He also quotes from several
books of the Bible, the Psalms (i. 8 and v. 6), 2 Corinthians
(i. 9), Malachi (i. 34), John (i. 38 and vi. 22), Joshua (vii. 1),
1 Kings (xi. 3), and Revelation (xiv. 2), some of these quotations being taken over from Bede. Beyond this, there are a
number of vague references to other men and their opinions
(iii. 6; vi. 1; vi. 5; vii. 1; viii. 11; ix. 1), but, with the possible exception of one, they do not indicate that Ælfric used
sources other than Bede's writings. The exception is the
allusion to 'priests', i.e. computists, in vii. 1, which to me at
least seems proof that in this place Ælfric joins issue with a
text repeatedly found in Old English computi.[3]

In this edition the attempt has been made to give the
sources of Ælfric's text as fully as possible and, wherever
practicable, paragraph by paragraph. It has long been known
that Bede's treatises *De Temporum Ratione, De Temporibus*,
and *De Natura Rerum* furnished Ælfric's materials. The table
of correspondences given by Reum[4] fails to show their rela-

---

[1] Cp. Edward Schröder, 'Einiges vom Buchtitel in der englischen Literatur des Mittelalters', *Anglia*, lxii (1938), p. 248.

[2] Bede in this place mentions, and quotes from, *De Trinitate*. Even
though Ælfric knew this book (his homily *De Fide Catholica*, Thorpe, i.
274, is based on it), there is no reason to assume that he used it when compiling his *De Temporibus Anni*.

[3] My reasons for believing this are found on p. 96, below.

[4] 'De Temporibus. Ein echtes Werk des Abtes Ælfric', *Anglia*, xi. 468.

tive importance, as it is incomplete and inaccurate. Reference to the sources as printed in this edition will show that Ælfric drew much more heavily on *De Temporum Ratione* than on the two shorter books. I have also found that Ælfric had other sources beside these three, although they are of subsidiary importance. The introductory portion on the Creation is based directly upon the Bible, and here as well as in some later places Ælfric shows that he knew and used Bede's *Commentarii in Principium Genesis*.[1] Furthermore, it appears that Ælfric knew Isidore's *De Natura Rerum*, but since Bede in his turn often agrees with Isidore it is difficult to say which of the two Ælfric may have had before him. There are, however, several passages in Ælfric for which I cannot find the source in Bede, but that can be paralleled from Isidore. Finally, the work entitled *De Ratione Computi Liber*, which is printed under the name of Bede in Migne's *Patrologia Latina*, xc. 579–600,[2] contains much that could have been quoted as source-material for Ælfric. It is written in dialogue form and appears to be an extract from *De Temporum Ratione*. Its presentation of the facts is often more direct than that of the longer work, and thus it agrees in places more closely with Ælfric than does *De Temporum Ratione*. But since it offers nothing that cannot be found in the latter, and since it cannot be proved that Ælfric knew it, I have refrained from using it in compiling the sources.

The term 'source' must be understood with some latitude. Ælfric uses his authorities, he does not translate them consecutively. Only occasionally does he translate a sentence or a short paragraph more or less closely. He also tends to simplify and to substitute brief statements of fact for theoretical considerations. This is done extremely well in most cases, but in a few places his treatment is so concise that the full meaning of his text becomes apparent only when com-

[1] Printed in vol. vii of Giles's edition.
[2] Giles (vol. vi, p. xiii) rejected it as spurious. C. W. Jones, *Bedae Pseudepigrapha*, Ithaca, 1939, p. 41, thinks the work was probably composed in the vicinity of Mainz before A.D. 771.

pared with the more exhaustive discussion of Bede (e.g. iii. 7, viii. 2–6). Ælfric omits Bede's calculations and his exposition of the mathematical foundation of astronomical phenomena. He eschews argument, which abounds in Bede's writings, in favour of a straightforward presentation of the latter's results. Exceptions are the passages where he repeats Bede's reasoning concerning the right date of the vernal equinox (vi. 1–6) and the distinction between the natural and the artificial day (i. 24). He also follows Bede when he denies that the stars fall from the skies (ix. 1), and when he distinguishes between the natural properties and influences (*effectiva potentia*) of the moon and what is falsely ascribed to her by superstitious men (viii. 11). On the other hand, he himself seems to have added the arguments regarding the right beginning of the year (iv. 21) and the origin of the bissextile year (vii. 1–2). The reference to differences of opinion as to the time when the moon should be accounted new (viii. 2–6) is based on Bede, but somewhat twisted. In all these cases, however, Ælfric is dealing with living issues, questions of practical importance to the computist that were either undecided or gave trouble to the less learned. He hardly ever states any reasons for deciding a problem this way or the other and thus is very much briefer than is Bede. He indulges nowhere in an argument purely for the sake of a scientific controversy.

Reum[1] collected the passages which, according to him, show Ælfric's predilection for examples taken from everyday life, and for concrete similes. The list is not very long, and several items, as has been shown,[2] do not count, as they can be traced in Ælfric's source. On the other hand, Ælfric omits regularly (with the exception of two brief passages: iv. 27 and x. 10–11) Bede's references to scientific experiment by which astronomical phenomena can be reproduced on earth.[3]

---

[1] Op. cit., p. 486. [2] See p. xliv, above.
[3] Cp. *DTR.* v (Giles, p. 152, middle); vii (Giles, p. 158, top); xvii (Giles, p. 183, middle); xxvi (Giles, p. 197 sq.); xxxi (Giles, p. 208, middle); xxxii *in fine* (Giles, p. 211); xxxiv *in fine* (Giles, p. 217).

Thus his treatise is somewhat hard and fast, it acquires a doctrinaire quality which is not altogether pleasing. Ælfric was no scientist. He knew his Bede well enough, but to him science was an auxiliary branch of learning, the handmaid of theology. From all this it follows that it is often difficult to say what passage in Bede should be considered Ælfric's source. I have given as the source whatever section from Bede's writings comes nearest to Ælfric's wording. There are, however, considerable variations in the closeness of these correspondences. Some gave Ælfric his material only, others gave him the words as well.

Under the general heading of 'Parallels' I have quoted (under the line on the right-hand pages): (1) parallel passages from Bede and occasionally from other Latin authors, (2) parallel passages from Ælfric's works (these are important in that they help establish his authorship of our tract), and (3) parallel passages from OE. literature, especially from Byrhtferð. With regard to these quotations from Byrhtferð it should be observed that some of them represent literal borrowings on the part of Byrhtferð, whilst others are adduced merely because they deal with the same subject-matter. In the latter case Bede is usually the common source of both the later authors. In comparing such passages the superior skill of Ælfric, both as a writer and as a teacher, becomes very apparent.[1]

§ 9. ARRANGEMENT OF THE PRESENT EDITION

(1) The text is based on MS. G. Departures from it are indicated in special footnotes. Expanded contractions are denoted by the use of italics, and words inserted by the editor are enclosed in square brackets.

(2) Proper names and the word *God* 'deus' have been capitalized by the editor. Where the MS. writes a word

---

[1] Cp., e.g., iii. 27-9.

partly in capitals, partly in minuscules, it is usually printed wholly in capitals, sometimes wholly in minuscules. The round letter *S* is used in the MS. forty-six times. It is not quite clear whether the scribe really meant a capital, or whether he used the character as an alternative to the insular *s*. Sometimes, but not always, he writes *S* after a period (which roughly corresponds to our comma), especially in words like *se, seo,* and *swa.* On the other hand, *S* is used in the middle of a sentence or a phrase, where there is no reason for a capital. This occurs repeatedly with the word *sunne* and in isolated cases with some other words. I have retained the capital in print where it is used in a Latin word or a proper name: *Saltus* viii. 1; *Septem* ix. 5; *Septemtrio* ix. 6 and x. 20; *Scipsteorra* ix. 8; *Subsolanus* x. 17. In all other cases I have printed small *s*.

(3) The division of the text into chapters and paragraphs is found in MS. G, but their numbering is the editor's.

(4) The apparatus criticus has all deviations of the other MSS. from MS. G with the exception of purely graphic variants. Different use (or lack) of capitals, of contractions, and of *þ* instead of *ð* (or vice versa) are not noted. No account is taken of the accents of the variant MSS. It must therefore neither be assumed that they have the accents of MS. G, nor that they may not have other or additional accents.

(5) The following abbreviations are used for the texts quoted as Sources and Parallels:

    a. DT., DTR., DNR. stand for Bede's *De Temporibus, De Temporum Ratione,* and *De Natura Rerum,* ed. Giles, vol. vi.

    b. Bede's *Commentarii in Librum Genesis* are quoted as *Bede (Giles vii).*

    c. Isidore, *DNR.* is his *De Natura Rerum.*

    d. Ælfric's *Interrogationes* are quoted from Tessmann's edition. The Roman numerals refer to the questions or answers.

e. Ælfric's *Hexameron* is quoted from Crawford's edition, by the line.

f. His *Lives of Saints* are quoted from Skeat's edition. The numerals refer to the number of the homily and to the line.

g. His *Genesis* and *Joshua* from Crawford's *Heptateuch*, by chapter and verse.

h. The following are quoted by the page: Ælfric's *Homilies* from Thorpe's edition; the OE. *Martyrologium* from Herzfeld; Byrhtferð's *Manual* from Crawford; Wulfstan's *Homilies* from Napier.

i. Occasional quotations from other authors are given with full reference.

DE TEMPORIBUS ANNI

DE TEMPORIBUS ANNI

Her æfter fyligð ân lytel cwyde be gêarlicum tidum · þæt nis
to spelle geteald · Ac elles to rædenne · þam ðe hit licað ·

[I. DE DIE.]

1. Ic wolde eac gif ic dorste gadrian sum gehwæde andgit of
ðære bec þe BEDA se snotera lâreow gesette · 7 gegaderode of manegra wisra lareowa bocum be ðæs geares
ymbrenum · fram anginne middaneardes;
2. Witodlice ða ða se ælmihtiga scyppend þisne middaneard gesceop · þa cwæð hê gewurðe leoht · 7 leoht wæs
ðærrihte geworden ;
3. Ða geseah God þæt þæt leoht wæs gôd · 7 todælde þæt
leoht fram ðam þeostrum · 7 het þæt leoht dæg · 7 ða þeostru
niht · 7 wæs ða geteald æfen 7 merigen to anum dæge ;
4. On ðam oðrum dæge gesceop God heofonan · seo þe is
gehâten firmamentum · seo is gesewenlic · 7 lichamlic · ac swa
ðeah we ne magon for ðære fyrlenan | heahnysse · 7 þæra

VARIANT READINGS

Title *in* DG *only.* B *has the* rubric: INCIPIUNT PAUCA DE
TEMPORIBUS BEDAE PRESBITERI.
Preamble *in* G *only.*
I. Chapter-heading DE DIE *omitted in all MSS.* AD *place chapter I at the end.* D *marks the seven days of the Creation in the margin
by Roman numerals, i.e. it has the numerals I–VII opposite the beginnings of* §§ *1, 4, 10, 12, 14, 15, 16.*
1. B *writes the first nine words in large red capitals.*   D. Ic wolde—
bec þe *desunt*   B. gyf   B. pluccian *for* gadrian   A. and-
gint, B. andgyt   B. gaderode   A. ymbrynum   B.
annginne   D. middangeardes   B *adds*: Ðæt nis to spelle ac
elles to rædenne þam þe hit licað
2. D. þa *for* ða ða   D. þysne middangeard   BD. geweorðe
3. AB. good   A. ðam *deest*   A. þystrum   A. þystru,
B. ðeostro   A. mergen, BD. merien
4. BD. heofenan   A. ys   A. ys   A. ferlenan, B. fyrly-
nan   A. heahnesse

SOURCES

2. In principio creauit Deus coelum et terram . . . Dixitque Deus: Fiat lux. Et facta est lux. *Genesis i. 1, 3.*

3. Et uidit Deus lucem quod esset bona: et diuisit lucem a tenebris. Appellauitque lucem Diem, et tenebras Noctem: factumque est uespere et mane, dies unus. *Genesis i. 4–5.*

4. Et fecit Deus firmamentum . . . Vocauitque Deus firmamentum, Coelum: et factum est uespere et mane, dies secundus. *Genesis i. 7–8.*

PARALLELS

2–3. *DTR. v passim.—DNR. ii.—Ælfric's Genesis i. 1, 3–5.*
2. He cwæð, 'Geweorðe leoht,' and ðærrihte wæs leoht geworden. *Homilies, i. 14.*
4. *DNR. ii.—Ælfric's Genesis i. 7–8.*—Micel swêg gæð . . . of þâm scînendan rodore, þêah þe wê for þâm mycclan fyrlene hit gefrêdan ne magon. *Interrogationes, xxi.*

wolcna þicnysse · 7 for ure eagena tyddernysse hî næfre
geseon;

5. Seo heofen belicð on hire bosme ealne middaneard · 7
heo æfre tyrnð onbuton us swyftre ðonne ænig mylenhweowul
· eal swa deop under þyssere eorðan swa heo is bufon;

6. Eall heo is sinewealt 7 ansund · 7 mid steorrum amett;

7. Soðlice ða oðre heofenan ðe bufon hire sind · 7 beneoðan ·
sind ungesewenlice · 7 mannum unasmeagendlice;

---

4 (*cont.*) A. þicnesse   A. eagen   A. tyddernesse   A. hyne
eac geseon *for* hi næfre geseon
5. B. heofon   ABD. belycð   ABD. hyre   D. middangeard
A. abutan, B. onbutan   AB. swiftre   A. þænne   A. ænigre
milne   B. mylnnhweol   BD. eall   A. þisse, B. þyssre, D.
þissere   A. ys   AB. bufan
6. A. midd   BD. amet
7. B. bufan   ABD hyre   A. syndon, BD. synd   A. 7
beneoðan sind *desunt*   BD. synd   A. gesewenlice, B. ungesegenlice

4–6. Coelum subtilis igneaeque naturae, rotundumque, et a centro terrae aequis spatiis undique collectum. Unde et conuexum mediumque quacunque cernatur, inenarrabili celeritate quotidie circumagi sapientes mundi dixerunt. *DNR. v.*

6. . . . de nostro coelo, in quo sunt posita luminaria. *Bede, Comm. in Lib. Gen. (Giles vii. 4)*.

7. Superius illud coelum, quod mortalium omnium est inaccessibile conspectibus. *Bede (Giles vii. 4)*.

7–9. Ambrosius sanctus, in libro Hexaemeron sic loquitur dicens: '. . . Philosophi autem mundi septem coelos, id est, planetarum globos consono motu introduxerant . . .' Siquidem in ecclesiasticis libris et 'coeli coelorum' leguntur, et apostolus Paulus usque ad tertium coelum fuisse se intelligit

---

4–5. On ðam oðrum dæge ure Drihten geworhte ðone firmamentum ðe men hatað rodor, se belycð on his bosme ealle eorðan bradnysse . . . 7 he æfre gæð abutan swa swa yrnende hweowol . . . se gæð under ðas eorðan eal swa deop swa bufan. *Hexameron*, 140–2, 144, 151.—On þæm dæge god gescop þone rodor betweoh heofone 7 eorðan . . . se rodor ymbfehð utan eall, sæ 7 eorðan. *Martyrologium, p. 40*.

4–8. On þissum monðe gesceop God Ælmihtig ealle gesceafta gesewenlice 7 ungesewenlice. He cwæð, 'Gewurðe leoht!' 7 hyt gewearð. Se dæg wæs on .XV. kalendas Aprelis. On þam oðrum dæge, He geworhte firmamentum, þæt ys þeos heofon. Heo ys gesewenlic 7 lichamlic, ac swaþeah we ne magon hig næfre geseon for þære fyrlenan heahnysse. Seo heofon belycð on hyre bosme ealne middaneard, 7 heo æfre tyrnð onbutan us. Heo ys swyftre þonne ænig mylenhwiol, eall swa deop under þisre eorðan swa heo ys bufan. Eall heo ys synewealt 7 ansund, 7 mid steorrum amet. Soðlice þa oðre heofenan, þe bufan hyre synt 7 beneoðan, synt ungesewenlice 7 mannum unasmeagendlice. Synd swa þeah ma heofena, swa swa se witega cwyð: Celi celorum. *Byrhtferð, p. 80*.

5. *Cp. v. 4–5 and ix. 9 below.*—He (*sc.* God Ælmihtig) besceawað þa niwelnyssa þe under þyssere eorðan sind. He awecð ealle duna mid anre handa, [and ealle eorðan he belycð on his handa]. *Homilies, i. 8–10 and Sisam, Review of English Studies, viii. 62 sq.*

5–6. Hwilces gecyndes is seo heofon? Fyrenes gecyndes and sinewealt and symle turnigende. *Interrogationes, xx.*

7. Ða upplican heofonas ða englas on wuniað, he geworhte eac ða on ðam ylcan dæge. *Hexameron, 118 f.*

8. Sind swa ðeah mâ heofenan · swa swa se witega cwæð;
Celi celorum · þæt is heofena · heofenan;
9. Eac se apostol Paulus awrât · þæt he wæs gelæd oð þa
ðriddan heofenan · 7 he ðær gehyrde ða digelan wôrd ðe nân
man sprecan ne môt;

10. On ðam ðriddan dæge gesceop se ælmihtiga God sæ · 7
eorðan · 7 ealle eorðlice spryttinga;
11. Þas ðry dagas wæron buton sunnan 7 monan · 7 steorran · 7 eallum tidum · gelicere wægan mid leohte · 7 þeostrum
aðenede;
12. On ðam feorðan dæge gesceop God twa miccle leoht ·
þæt is sunne 7 mona · 7 betæhte þæt mare leoht þæt is seo
sunne to ðam dæge · 7 þæt læsse leoht þæt is se mona to ðære
nihte;
13. On ðam ylcan dæge he geworhte ealle steorran · 7 tida
gesette;

---

8. A. Syndon, BD. Synd   AB. heofonan   BD. Cęli   B.
cælorum, D. cęlorum   A. þæt is desunt   B. his for is   A.
heofonan heofone, B. heofona heofonan
9. ABD. gelædd   A. on for oð   A. þryddan   AB.
heofonan   A. geherde   ABD. diglan   ABD. mann
10. A. þryddan   B. gescop   A. spyttinga
11. B. Ða   A. das for dagas   AB. butan   BD. steorrum
A. þystrum
12. A. On þam feorðan dæge he gesceod sunnan to þam dæge ·
7 þæt læsse leoht . . .   B. scop   D. micele   A. ys

raptum. Sed de numero eorum nihil sibi praesumat humana temeritas. *Isidore, DNR. xiii.*

8. Laudate eum coeli coelorum: et aquae omnes, quae super coelos sunt, Laudent nomen Domini. *Psalmus 148, 4-5.*

9. Scio hominem in Christo ante annos quatuordecim . . . raptum huiusmodi usque ad tertium coelum . . . et audiuit arcana uerba, quae non licet homini loqui. *Ad Corinthios ii. 12, 2-4.*

10. . . . tertio, species maris et terrae, cum eis quae terrae radicitus inhaerent. *DNR. ii.*

11. Nam praecedens triduum, ut omnibus uisum est, absque ullis horarum dimensionibus, utpote necdum factis sideribus, aequali lance lumen tenebrasque pendebat. *DTR. vi.*

12. Fecitque Deus duo luminaria magna, luminare maius, ut praeesset diei, et luminare minus, ut praeesset nocti, et stellas. *Genesis i. 16.*

13. Dixit autem Deus: Fiant luminaria in firmamento

---

8. Heriað hine heofonas ðæra heofona heofonas, and eac ða wæteru ðe bufan heofonas synd herian hi Godes naman; ðus segð ðæt halige gewrit. *Hexameron, 159-61.*

9. Men ða leofostan, Paulus se Apostol, ealra ðeoda lâreow, awrât be him sylfum þæt hê wære gelædd up to heofonum, oð þæt hê becom to ðære ðriddan heofonan; and hê wæs gelæd to neorxna-wânge, and þær ða gastlican dygelnysse gehyrde and geseah; ac hê ne cydde na eorðlicum mannum, ðaða hê ongean com, hwæt hê gehyrde oððe gesawe, ðisum wordum writende be him sylfum . . .: 'Ic wât ðone mann on Christe, þe wæs gegripen nu for feowertyne gearum, and gelæd oð ða þriddan heofenan; and eft hê wæs gelæd to neorxna-wange, and ðær gehyrde ða digelan word þe nân eorðlic mann sprecan ne môt.' *Homilies, ii. 332. Cp. also Homilies, i. 392.*

10-14. *DNR. ii.*

12-13. He geworhte twa mycle leohtfatu, swa Genesis geseð, þæt ys seo forme boc on biblitheca. He gesceop sunnan 7 monan, 7 tungla, 7 steorran. *Byrhtferð, p. 8. Cp. ibid., p. 82.*—God geworhte ða sona twa scinende leoht, mycele 7 mære, monan and sunnan, ða sunnan on merigen to ðæs dæges lihtinge, ðone monan on æfen mannum to lihtinge on nihtlicere tide mid getacnungum. 7 ealle steorran he eac ða geworhte. *Hexameron, 206-11.*

12-14. *Ælfric's Genesis i. 16, 14, 20-1.*

14. On ðam fiftan dæge he gesceop eal wyrmcynn · 7 ða micclan hwalas 7 eal fisccynn · on mislicum 7 menigfealdum hiwum;

15. On ðam sixtan dæge he gesceop eal deorcynn · 7 ealle nytenu þe on feower fotum gâð · 7 þa twegen men Adâm · 7 Euan;

16. On ðam seofoðan dæge hê geendode his weorc · 7 seo wucu wæs ða agân;

17. Nu is ælc dæg on ðisum middanearde of ðære sunnan lihtinge;

18. Soðlice seo sunne gæð be Godes dihte betwux heofenan 7 eorðan · on dæg bufon eorðan · 7 on niht under ðysse eorðan eal swa feorr adûne on nihtlicere tide under þære eorðan · swa heo on dæg bufon upastihð;

---

14. B. gescop   ABD. eall   D. fugel- *for* wyrm-   AD. hwælas   ABD. eall   B. mistlicum   A. manig-, B. mænig-
15. ABD. syxtan   B. gescop   ABD. eall   B. nytena D. gað 7 eall wyrmcynn   ABD. menn   B. efan
16. A. hys
17. A. ys   A. ælce   B. ðysum, D. þysum   D. middangearde   B. lyhtinge
18. A. mihte 7 dihte   B. betweox   A. heofonan   AB. bufan   A. ðysse *deest*, D. þissere   BD. eall   A. feower *for* feorr, D. feor   A. nihtricre, BD. nihtlicre   D. heo *deest* ABD. bufan   A. upp-

coeli, et diuidant diem ac noctem et sint in signa et tempora
et dies et annos. *Genesis i. 14.*

14. Dixit etiam Deus: Producant aquae reptile animae
uiuentis, et uolatile super terram sub firmamento coeli!
Creauitque Deus cete grandia et omnem animam uiuentem
atque motabilem quam produxerant aquae in species suas,
et omne uolatile secundum genus suum. *Genesis i. 20–1.*

15. Et fecit Deus bestias terrae iuxta species suas et
iumenta et omne reptile terrae in genere suo . . . Et creauit
Deus hominem ad imaginem suam, . . . masculum et feminam
creauit eos. *Genesis i. 25, 27.*

16. Compleuitque Deus die septimo opus suum. *Genesis
ii. 2.*

17. Dies est aer sole illustratus. *DTR. v.*

18. . . . qui (*sc.* sol) secum semper et ubique lumen diurnum
circumferens, non minore aerum spatio noctu subter terras
quam supra terras interdiu creditur exaltari. *DTR. v.*

---

14. On þam fiftan dæge, þæt ys on .XI. kalendas, He gescop eall
wyrmcynn 7 creopende 7 fleogende 7 swymmende 7 slincgende 7 þa
myclan hwælas 7 þa lytlan sprottas 7 eall fisckynn on myslicum 7
mænigfealdum hiwum. *Byrhtferð, p. 82.*—On ðam fiftan dæge ure
Drihten gesceop of wætere anum ealle fixas on sǽ 7 on éauum 7 eall
ðæt on him crypð 7 ða mycelan hwalas on heora cynrynum. *Hexameron, 239–42.*

15–16. *Ælfric's Genesis i. 25, 27; ii. 2.*—On þam syxtan dæge, þæt
ys on .X. kalendas Aprelis, He gescop eall deorcynn 7 ealle nytenu
þe on feower fotum gáð 7 þæne man, Adam 7 Euan, 7 þa He gebletsode. On þam seofoðan dæge, He geendode His weorc, þæt ys .IX.
kalendas Aprelis, 7 seo wucu wæs agán, 7 He gebletsode þæne dæg.
Byrhtferð, p. 82.

15. Hê ða gesceop of ðære eorðan eall nytencynn, and deorcynn,
ealle ða ðe on feower fotum gað. *Homilies, i. 16.*

16. On ðam seofoðan dæge hê geendode his weorc. *Homilies, i. 14.*

17. Forþon ys dæg gecweden, forþon þæt lyft byð aliht þurh þære
sunnan leoman. *Byrhtferð, p. 112.*

18. Heo (*sc.* seo sunne) is gesceaft, and gæð be Godes dihte. *Homilies, i. 288.* Cp. *ibid., i. 172, and xi. 10 below.*

18, 21. Talis ergo schematis terra mortalibus ad inhabitandum
data, solis circuitus in hoc mundo lucentis certa ratione constitutionis Dei, alibi diem exhibet, alibi noctem relinquit. *DTR. xxxii.*

19. Æfre heo bið yrnende ymbe ðas eorðan · 7 eal swa leohte scinð under ðære eorðan on nihtlicere tide · swa swa heo on dæg deð · bufon urum heafdum;

20. On ða healfe ðe heo scinð þær bið dæg · 7 on ða healfe ðe heo ne scinð þær bið niht;

21. Æfre bið on sumere sidan þære eorðan dæg · 7 æfre on sumere sidan niht;

22. Þæt leoht ðe we hatað dægered · cymð of ðære sunnan · þonne heo upweard bið · 7 heo ðonne todræfð þa nihtlican ðeostru ðurh hire micclan leohte;

23. Eal swa ðicce is seo heofen mid steorrum afylled on dæg swa on niht · ac hî nabbað nâne lihtinge for ðære sunnan andwerdnysse;

24. We hatað ænne dæg fram sunnan upgange oð æfen · ac

f. 256 r.   swa ðeah on bocum is | geteald to anum dæge fram ðære sunnan upgange oð þæt heo eft becume þær heo ær uppstah · on ðam fæce sind getealde feower 7 twentig tida;

The first fragment of H *begins here.*

19. H. Efre   H. seo sunne *for* heo   A. byþ, B. byð   H.
ymb   BDH. eall   H. þare   ABDH. nihtlicre   BDH.
bufan   H. urum heafdum *desunt*
20. A. byþ, B. byð   H. 'he'o   B. byþ
21. A. byþ, B. byð   BH. sumre   H. þare   AH. æfre
deest   BH. sumre
The first fragment of H *ends here.*
22. A. dægræd, BD. dægred   A. on *for* of   A. þænne
A. byþ   A. þænne   A. þysru   ABD. mid *for* ðurh   ABD.
hyre
23. BD. Eall   A. ys   B. þeo   B. heofon   A. hig
A. nanne   B. lyhtinge   AD. andweardnysse
24. A. anne   A. ys   B. is on bocum geteald   D. heo
becume eft   BD. up-   AD. uppastah   A. syndon, BD. synd

19–21. . . . eo tempore quo nox apud nos est, eas partes mundi praesentia lucis illustret, per quas sol ab occasu in ortum redit: ac per hoc omnibus XXIV horis non deesse per circuitum gyri totius alibi diem, alibi noctem. *DTR. v.*

22. Diluculum, quasi iam incipiens diei lux, haec et Aurora, solem praecedens. *DTR. iii.*

23. . . . easque (*sc.* stellas) diei aduentu celari. *DNR. xi.*

24. Et uocauit Deus lucem diem. Quae definitio bifariam diuiditur, hoc est, uulgariter et proprie. Vulgus enim omnem diem solis praesentiam super terras appellat. Proprie autem dies XXIV horis, id est, circuitu solis totum orbem lustrantis impletur . . . Quod multorum quidem et nostrorum et secularium literarum testatur auctoritas. *DTR. v.*

---

19. Eall swa gæð seo sunne 7 soðlice se mona abutan ðas eorðan mid bradum ymbhwyrfte eall swa feorr beneoðan swa swa hi bufan ús gað. *Hexameron, 236–8.*

19–21. Seo sunne æfre byð yrnende ymbe þas eorðan, 7 eal swa leohte heo scinð under þære eorðan on middre nihte, swa heo deð bufan eorðan on middæge. On þa healfe þe heo scinð þær byð dæg; 7 on þa healfe þe heo ne scinð þær byð niht. Æfre byð on sumere sidan þære eorðan dæg, 7 on sumere sidan niht. *Byrhtferð, p. 124.*

22. Heo (*sc.* seo sunne) totwæmð þære nihte þystru mid hyre beorhtnysse. *Byrhtferð, p. 112.*—Dægred todræfð þa dimlican þystra and manna eagan onlyht þe blinde wæron on niht. *Lives, v. 108.*—Se beorhta dæg todræfð þa dimlican þeostru ðære sweartan nihte. *Homilies, i. 604.*

24. On twam wisum ys se dæg gecweden, naturaliter et uulgariter, þæt ys gecyndelice 7 ceorlice. Þæt ys þæs dæges gecynd, þæt he hæbbe feower 7 twentig tida fram þære sunnan upsprunge þæt he eft up hyre leoman ætywe. Vulgaris uel artificialis dies est, þæt byð ceorlisc dæg oððe cræftlic, fram þære sunnan anginne þæt heo to setle gá, 7 eft cume mancynne to blisse. *Byrhtferð, p. 112.*

25. Seo sunne is swiðe micel · eal swa brâd heo is þæs ðe
bêc secgað swa eal eorðan ymbhwyrft · ac heo ðincð ûs swiðe
unbrâd · forðan ðe heo is swiðe feor fram urum gesihðum;
26. Ælc ðing swa hit ðe fyrr bið · swa hit ðe læsse ðincð;
27. We magon hwæðere tocnâwan be hire leoman · *þæt*
heo unlytel is;
28. Swa hraðe swa heo upastihð heo scinð geond ealle
eorðan gelice · 7 ealre eorðan brâdnysse endemes oferwrihð;
29. Eac swilce ða steorran ðe us lytle ðincað · sind swiðe
brade · ac for ðam micclum fæce þe us betweonan is hî sind
geðuhte urum gesihðum swiðe gehwæde;
30. Hi ne mihton swa ðeah nân leoht to eorðan asendan ·
fram ðære healican heofenan · gif hî swa gehwæde wæron swa
swa urum eagum ðincð;
31. Soðlice se môna 7 ealle steorran underfoð leoht of
ðære micclan sunnan · 7 heora nân næfð nænne leoman buton
of ðære sunnan leoman;

32. 7 ðeah ðe seo sunne under eorðan on nihtlicere tide
scine · þeah astihð hire leoht on sumere sidan þære eorðan

---

25. A. ys   A. micel · 7 heo ys eall swa brad   B. mycel
AB. eall   A. secgeaþ eall swa   ABD. eall   D. ymbhyrft
B. þingð   B. swyðe   ABD. forþam   A. ys   ABD.
feorr
26. A. þinge, B. þingð   B. þe hit *for* swa hit ðe   A. fyr
A. byþ, B. byð   B. þe *for* swa   B. ðingð, D. þingð
27. ABD. magon þeah hwæþere   A. tocnapan   ABD. hyre
A. ys
28. B. raðe   A. upp-   A. eond *for* geond   A. endemes
heo oferwrihð   B. oferwryhð
29. B. swylce   A. litle, B. lyttle   A. þinceaþ, B. þingeað
A. sindon, BD. synd   B. swyðe   B. 7 *for* ac   AD. micclan,
B. miclum   A. betwynan   AB. ys   A. syndon, BD. synd
A. swyþe
30. A. hig   D. on *for* to   A. heofonan   B. Gyf   A.
hig   B. wæro   A. eo *for* urum   B. ðincð *deest*
31. B. miclan   A. hyra   A. nanne   A. butan   D.
hyre *for* ðære sunnan
32. ABD. nihtlicre   D. scinð   ABD. hyre   B. sumre

25. Argumentantur solem terra esse maiorem, quamuis ob immensam longinquitatem modicus uideatur. *DTR. vii.*

26. Omnia enim longius posita solent breuiora uideri. *Bede, Comm. in Lib. Gen. (Giles vii. 16).*

31. Luminare maius est sol, non solum forma sui qualiscumque est corporis, sed etiam magnitudine luminis, qua et ipsum luminare minus et stellas illustrare creditur. *Bede, Comm. in Lib. Gen. (Giles vii. 16).*—Stellae lumen a sole mutuantes. *DNR. xi.*

32. Sed et ortus sol lunam stellasque maiore lumine, ne terram inluminent, impedit. *Bede (Giles vii. 15).*

---

25. Solis ignem dicunt aqua nutriri, multoque hunc luna ampliorem : lunam uero terra esse maiorem, unde et cunctis unius magnitudinis apparet. Quod enim nobis quasi cubitalis uidetur nimiae celsitudinis distantia facit. *DNR. xix.*

26. Omnia enim quae proxima sunt nobis maiora uidentur, longinquitate autem locorum uisus languescit. *Isidore, DNR. xvi. 3. Cp. ibid. xxii. 2.*

28. Swa hraðe swa heo (*sc.* seo sunne) upasprincð on ærne merigen, heo scinð on Hierusalem, and on Romebyrig, and on ðisum earde, and on eallum eardum ætgædere. *Homilies, i. 286.*

þe ða steorran bufon us onliht · 7 ðonne heo upagǣð · heo
oferswið ealra ðæra steorrena · 7 eac þæs monań leoht · mid
hire ormǣtan leohte;

33. Seo sunne getacnað urne hælend Crist · se ðe is rihtwisnysse sunne · swa swa se witega cwæð;

34. Timentibus autem nomen domini · orietur sol iustitię ·
et sanitas in pennis eius; Þam mannum þe him ondrædað
Godes naman · ðam arist rihtwisnysse sunne · 7 hælð is on
hire fiðerum;

35. Se mona ðe weaxð 7 wanað getacnað þas andwerdan
gelaðunge · ðe we on sind;

36. Seo is weaxende þurh acennedum cildum · 7 wanigende
þurh forðfarendum;

37. Ða beorhtan steorran getacniað ða geleaffullan on
Godes gelaðunge · ðe on goddre drohtnunge scinað;

38. Crist soðlice onliht hî ealle þurh his gife · swa swa se
godspellere Iohannes cwæð;

39. Erat lux uera · quę inluminat omnem hominem
uenientem in hunc mundum; Þæt soðe leoht com þe onliht
ælcne mannan · cumendne to ðisum middanearde;

---

32 (*cont.*) AB. bufan   A. onlihtaþ   A. þænne   A. gæþ
for upagǣð   A. steorrena leoht 7   A. leoht *deest*   ABD. hyre
33. AB. ys   A. rihtwisnesse
34. A. rihtwisnesse   A. hæl, B. hælþe, D. hælðe   A. ys,
D. is *deest*   A. his *for* hire, BD. hyre
35. A. wyxþ, D. weaxeð   A. 7weardan, BD. andweardan   A.
gewanunge *for* gelaðunge   BD. synd   A. ðe we on sind—
wanigende *desunt*
36. B. ys   D. acennede cild   BD. waniende   AD. forþ-
farende, B. forþfarenum
37. A. beorhtran   A. geleafullan   B. godre
38. B. onlyht   A. hig   A. hys   AB. gyfe   D. swa
for swa swa   A. Iohannes *deest*
39. A. Eact luix uera quem luminat   B. que   B. inluminet
B. onlyht   BD. mann   AD. cumende   B. ðysum, D.
þysum   A. earde *for* middanearde, D. middangearde

33–34. At uero iuxta spiritualem intelligentiam, sol Christus est, sicut in Malachia scriptum est: 'Vobis autem credentibus iustitiae sol orietur, et sanitas in pinnis eius'. *Isidore, DNR. xv. 3.*
34. *Malachia iv. 2.*
33–40. Oportet enim, ut sicut tunc primo sol potestatem diei, deinde luna cum stellis potestatem noctis accepit, ita et nunc ad insinuandum nostrae redemptionis gaudium, primo dies noctem longitudine adaequet, ac deinde luna plenissima eam luce perfundat, certi utique mysterii gratia: quia uidelicet sol ille creatus omnium illuminator astrorum, aeternam ueramque lucem significat, quae illuminat omnem hominem uenientem in hunc mundum. Luna autem et stellae, quae non proprio ut dicunt, sed aduentitio, et a sole mutuato lumine fulgent, ipsum ecclesiae corpus, et quosque uiritim sanctos insinuant. *DTR. vi.*
39. *Iohannes i. 9.*

---

33. Dominus Iesus Christus, sol æternus. *Isidore, DNR. xvii. 5.*—
Bede, Hist. Eccl. v. 22.—DTR. lxiv.—DT. xv.—Drihten, se ðe is
rihtwisnysse sunne. *Homilies, i. 144.*
35. Se móna hæfð þissere worulde gelicnysse, forðan ðe hê is
hwîltidum weaxende hwîltidum wanigende, swa swa ðeos woruld.
Homilies, ii. 214.—Heofonan rice getâcnað þas andwerdan gelaðunge,
forðan þe rihtwisra manna gegaderung is gecweden heofenan rice.
Ibid. ii. 72.
39. [Deus] . . . qui cum ipse sit lux uera. *Bede (Giles vii. 7).*—
Erat enim lumen uerum, non quod illuminatur, ut homo, sed quod
illuminat omnem hominem. *Augustine, Enarratio in Psalmum cxviii,
Sermo xxiii (Migne xxxvii. 1566).*—'Hwî gesceôp god lêoht on þone
forman dæg?' 'Hit gedafenode rihtlîce godes weorcum, þæt se
ælmihtiga, sê þe is êce lêoht, ǽrest þæt hwîlwendlîce lêoht geworhte,
þæt þâ ôþre gesceafta, þê hê syþþan gesceôp, mihton bêon þurh þæt
lêoht gesewene. God sylf is sôð lêoht, ac þæt gesceapene lêoht
geswutelode his weorc.' *Interrogationes, xxiv.—Cp. Homilies, i. 282,
286 and 614, and Lives of Saints, i. 71–6.*

40. Næfð ure nân nân leoht ænigre gôdnysse buton of
Cristes gife · se ðe is soðre rihtwisnysse sunne gehaten ;

[II.] DE PRIMO DIE SÆCULI · SIUE DE EQUINOCTIO
UERNALI.

1. Done forman dæg þyssere worulde we magon afindan
þurh ðæs lenctenlices emnihtes dæge · forðan ðe
se emnihtes dæg is se feorða dæg þissere worulde gesceapennysse ;
2. Ðry dagas wæron ær ðam dæge buton sunnan · 7 monan ·
7 eallum steorrum · 7 on þam feorðan dæge þyssere worulde
gesceapennysse gesceop | se ælmihtiga scyppend sunnan · 7
gesette hî on ærne merigen on middan eastdæle · þær ðæ[s]¹
emnihtes circul is geteald · þæt heo æfre ymbe geares ymbrynum þær ðone dæg · 7 ða niht geemnytte on gelicere wægan ;

f. 256 v.

¹ G. ðær

---

40. A. butan   AB. gyfe   AB. ys   A. rihtwisnesse   D
adds the colophon: þam sy wuldor 7 lof mid fæder · 7 halgan gaste ·
on ealra worulda woruld a butan ende · amen:–
II. *C begins here*   C. *title deest*   AD. Seculi
1. AD. þissere   C. þes   A. lentenlices, B. lænctenlices, D.
lenctenes   B. ymnihtes, D. emnyhtes   ABCD. dæg   ABD.
forþam   D. emnyhtes   A. ys   A. dæg *deest*   B. þysse,
C. þyssere   B. gescapennysse
2. ABC. butan   C. eallum *deest*   C. feorðam   A. þyss, D. þissere
A. gescepennysse, B. gescapennysse   C. þyssere worulde gesceapennysse *desunt*   A. hig   B. mergen, C. morgen   A. þes, G.
ðær   D. emnyhtes   C. ymbrenum   B. geymnytte ·   C.
gelicre   C. wæge

1. Bene quidem inquirentes . . . tempus aequinoctii non primo diei quo lux, sed quarto quo luminaria sunt facta, potius adsignarent. *DTR. vi.*

2. Nam praecedens triduum, ut omnibus uisum est, absque ullis horarum dimensionibus, utpote necdum factis sideribus, aequali lance lumen tenebrasque pendebat: et quarto demum mane sol a medio procedens orientis . . . aequinoctium quod annuatim seruaretur inchoavit: positus uidelicet cum primo

40. Swa eac ðurh Cristes tocyme to ðyssere worulde, and ðurh his apostola bodunge, tofeollon ða wiðerweardan weallas þyssere worulde ungeleaffulnysse. *Homilies, ii. 216.*—Hêr asette se apostol niht for ðære ealdan nytennysse, ðe rixode geond ealne middangeard ǽr Cristes to-cyme; ac he toscoc ða dwollican nytennysse ðurh onlihtinge his andwerdnysse . . . Deofol is eac niht gecweden, and Crist dæg, seðe us mildheortlice fram deofles ðeostrum alysde, and us forgeaf leoht ingehydes and soðfæstnysse. *Homilies, i. 602-4.*

1-3. On þam feorðan dæge, þæt ys on .XII. kł. Aprelis, He gesceop sunnan 7 monan 7 steorran 7 ealle tungla, 7 on ærne mergen þæs dæges up arás seo beorhte sunne riht on eastende þære heofone, 7 þæne monan þæs ylcan æfenes He gesette on þære ylcan stowe, 7 he wæs full swa swa he byð þonne he byð fiftyne nihta eald. Þæne forman dæg þisre worulde man mæg findan, swa ic herbufan cwæð, þurh þæs lengtenlican emnihtes dæg, forþon þe emnihtes dæg ys se feorða dæg þissere worulde gesceapennysse; and se lareow Beda telð mid micclum gesceade þæt se dæg is .XII. Kł Aprilis. *Homilies, i. 100.*

1-4. Se eahteteoða dæg þæs monðes þe we hâtað Martius, ðone ge hatað Hlyda, wæs se forma dæg ðyssere worulde. On ðam dæge worhte God leoht, and merigen, and æfen. Ða eôdon þry dagas forð buton tîda gemetum; forðan þe tunglan næron gesceapene, ær on þam feorðan dæge. On ðam feorðan dæge gesette se Ælmihtiga ealle tungla and gearlice tîda, and hêt þæt hî wǽron to tâcne dagum and gearum. Nu ongynnað þa Ebreiscan heora geares anginn on þam dæge þe ealle tida gesette wæron, þæt is on ðam feorðan dæge woruldlicere gesceapenysse; and se lareow Beda telð mid micclum gesceade þæt se dæg is .XII. KĪ Aprilis. *Homilies, i. 100.*

2. *Cp. i. 11 above.*

3. Ðæs ylcan dæges hê gesette ðone mônan fulne on
æfnunge · on eastdæle mid scinendum steorrum samod · on
ðæs hærfestlican emnihtes ryne · 7 ða easterlican tid þurh
ðæs monan anginn gesette;

4. We willað furðor ymbe ðas emnihte swiðor sprecan on
gedafenlicere stowe · 7 we secgað nu sceortlice *þæt* se forma
dæg ðyssere worulde is geteald to ðam dæge þe we hâtað
Quinta decima k*alendas* aprilis · 7 ðæs emnihtes dæg is
gehæfd · swâ swâ Beda tæcð · ðæs on ðam feorðan dæge ·
þæt is · on duodecima k*alendas* aprel*is* ;

5. Embe ðis we sprecað eft swiðor · swa swa we ǽr behêton :

[III.] DE NOCTE.

1. Niht is gesett mannum to reste · on ðisum middanearde ;

2. Soðlice on ðam heofenlicum eðele nis nân niht gehæfd ·
ac ðær is singal leoht · buton ælcum ðeostrum ;

---

3. A. þæne   D. fullne   A. steorran   A. hærfæst-   B.
ymnihtes, D. emnyhtes   C. monan *deest*   A. angin, B. angynn
C. anginn he gesette
4. C. We willað—stowe 7 *desunt*   D. wyllað   D. emnyhte
ABD. gedafenlicre   A. secgeað   C. ðe   A. þisse, D. þissere
D. emnyhtes   B. ys   C. þæt *for* ðæs   ABCD. aprilis
5. C. *the whole paragraph deest*   A. ymbe   A. swyþor

III. AC. *title deest*

1. C. geset   BD. þysum   D. middangearde
2. ACD. heofenlican   A. nys   D. geteald *for* gehæfd   A.
æc   A. ys   AC. butan   A. þisrum, BC. þystrum

terris oriretur in eo coeli loco, quem philosophi quartam
partem Arietis appellant, eundemque peracto annuo circuitu . . . repetiturus, . . . ut ipse uernalis aequinoctii punctus
modo mane, modo meridie, modo uespere, modo medio noctis
occurrat. *DTR. vi.*

3. Luna e contrario uespere plenissima ; neque enim quid
imperfectum creator aequissimus instituit, stellis una ful-
gentibus, ac medio apparens orientis, quartam partem librae,
qua aequinoctium autumnale adseuerant, tenuit, initiumque
paschae suo consecrauit exortu. *DTR. vi.*

4. De quibus singulis suo loco consequentius, prout Dominus dederit, exponemus: nunc admonere contenti, XII
Calendarum Aprilium die occursum aequinoctii, et ante
triduum, hoc est, XV Calendarum earundem primum seculi
diem esse notandum. *DTR. vi.*

1, 3. Nox est solis absentia terrarum umbra conditi, mortalibus ad requiem facta. *DT. iii.*

2. Merito itaque coelum coeli non inane uel uacuum esse
factum memoratur, sed nec tenebris in eo uel abysso locus

---

3. *Cp. viii. 2 below.*
4. Forðanðe he on lenctentid, swa swa lareowas secgað, gesceop
ðone forman dæg ðyssere worulde, ðæt is on gerimcræfte XV kalendas
aprilis. *Hexameron, 114–16.*—Prima die, id est .XV. kalendas Aprilis,
facta est lux. *Byrhtferð, p. 212.*—On þone nigonteogðan dæg [þæs]
monðes bið se æftera worolde dæg. *Martyrologium, p. 38 (The leaf*
on which the primus dies must have been treated is wanting).
5. *Cp. vi. 1–7 below.*

1 Est autem noctis umbra mortalibus ad requiem corporis data.
DTR. vii.—Nu be þære nihte uton sum þing her uparæran. Heo ys
gesett manncynne to reste. *Byrhtferð, p. 122.*
2. On ecne gefean heofona rîces : ðær næfre leofe ne gedælað, ne laðe
ne gemetað, ne næfre dæg æfter dæge, ne niht æfter nihte, ac þær bið
êce bliss and ece wuldor and ece gefea. *Wulfstan, p. 204 sq.*

3. Ure eorðlice niht soðlice cymð þurh ðære eorðan sceade ;

4. Þonne seo sunne gǣð on æfnunge under ðyssere eorðan ·
þonne bið þære eorðan bradnys betwux ûs · 7 ðære sunnan ·
þæt we hire leoman lihtinge nabbað · oð þæt heo eft on oðerne
ende upastihð ;

5. Witodlice ðeah ðe hit wunderlic ðince · nis ðeos woruldlice niht nân ðing buton ðære eorðan sceadu betwux ðære
sunnan 7 mancynne ;

6. Woruldlice ûðwitan sædon þæt seo sceadu astihð up oð
þæt heo becymð to þære lyfte ufweardan · 7 ðonne beyrnð se
mona hwiltidum þonne he full bið on ðære sceade ufwearde ·
7 fagetteð oþþe mid ealle asweartað · forðan ðe hê næfð þære
sunnan leoht ða hwile ðe hê þære sceade ord oferyrnð · oð
þæt ðære sunnan leoman hine eft onlihton ;

7. Se môna næfð nan leoht buton of ðære sunnan leoman ·
7 he is ealra tungla nyðemest · 7 forði beyrnð on þære eorðan
sceade · þonne hê full bið · na symle swa ðeah for ðam bradan
circule þe is zodiacus gehâten · under ðam circule yrnð seo
sunne · 7 se mona · 7 ða twelf tunglena tacna ;

---

3. B. nyht   C. cumð   C. eorðe
4. AD. þissere   A. byþ, BC. byð   C. þare   A. bradnyss
B. betweox   C. þare   ABCD. hyre   B. lyhtinge   A.
uppastihð
5. ACD. wundorlic   B. þinge   A. nys   AC. butan   C.
þare   B. betweox   C. þare   B. mannkynne, C. mankynne
6. A. Weoruld-   BC. sædan   C. ðeo   A. upp   C.
beocumð   A. lifte, C. lyte   B. ufeweardan, C. ufewearde
A. becyrþ *for* beyrnð   A. hwyltidum   C. ful   A. byþ,
B. byð   B. ufeweardre, C. ufewearde   B. faggeteð   ABD.
forþam   A. sunnan *deest*   A. hwyle   A. leoma   A.
onlihteþ, C. onlihteð
7. C. butan   A. for *for* forði   C. he yrnð *for* beyrnð   C.
ful   A. byþ, BC. byð   A. simble, C. æfre *for* symle   D.
under þone circul   A. yrþ   C. þeo   B. ða *deest*   AD.
tunglena *deest*, C. tungla

remanere ullus perhibetur, quod Dominus Deus illuminat, et cuius lucerna est Agnus. *Bede (Giles vii. 4 sq.)*.

3–4. Est autem nox absentia solis terrarum umbra conditi, donec ab occasu redeat ad exortum. *DTR. vii.*

5. Neque enim aliud esse noctem, quam terrae umbram. *DTR. xxvii.*

6. . . . quam uidelicet umbram noctis ad aeris usque et aetheris confinium philosophi dicunt exaltari, et acuminatis instar pyramidum tenebris lunam, quae infima planetarum currit, aliquando contingi, atque obscurari. *DTR. vii.*

7. Dicunt enim eam (*sc.* lunam) . . . non habere lumen proprium, sed a sole illustrari. *DTR. xxv.*—Lunam uero aiunt, cum infimas sui circuli apsidas plena petierit, nonnunquam umbra memorata fuscari, donec paulatim centrum

---

5. Nû nis yfel nân þing, bûton gôdnysse forgǣgednyss, swâ swâ þêostru ne synd nân þing, bûton lêohtes forlǣtennyss. *Interrogationes, xlvii.*

6. We rædað on tungelcræfte þæt seo sunne bið hwiltidum þurh ðæs monelican trendles underscyte aðystrod, and eac se fulla môna færlice fagettað, þonne he ðæs sunlican leohtes bedæled bið ðurh ðære eorðan sceadwunge. *Homilies, i. 608.*

7. Sed ne singulis mensibus eclipsis fieret, latitudo signiferi lunam superius inferiusue transmittit. *DNR. xxii.*—Stati autem, atque non menstrui sunt utrique defectus, propter obliquitatem Signiferi, lunaeque multiuagos, ut dictum est, flexus, non semper in scrupulis partium congruente siderum motu. *DTR. xxvii.*—Se seofoþa is se môna, ealra tungla nyþemest. *Interrogationes, xxi.*—Cp. *iv. 26 below.*

7–10. Symble þonne se mona gangeð æfter þære sunnan, þonne weaxeð his leoht, ond swa he bið þære sunnan neár swa bið his leoht læsse, ond swa he bið hyre fyrr swa bið his leoht mﬞáre, ond hwæðre he bið symble þurh þa sunnan onlyhted. *Martyrologium, pp. 42–4.*

8. Witodlice þæs mônan trendel is symle gehâl 7 ansund ·
ðeah ðe hê eal endemes eallunge ne scine;

9. Dæghwomlice þæs monan leoht bið weaxende oððe
wanigende feower pricon · ðurh ðære sunnan leoman;

10. 7 he gæ̂ð dæghwomlice oþþe to ðære sunnan · oððe
fram ðære sunnan swa fela pricon · na þæt hê becume to ðære
sunnan · forðan ðe seo sunne is micele ufor þonne se mona sy;
11. He cymð swa ðeah foron ongean þære sunnan · ðonne
hê of hire ontend bið;
12. Symle hê went his hricg to ðære sunnan · þæt is se
sinewealta ende þe þær onliht bið;

f. 257 r. 13. We | cweðað þonne nîwne monan æfter menniscum
gewunan · ac hê is æfre se ylca ðeah þe his leoht gelomlice
hweorfe;
14. Þæt æ̂mtige fæc bufon ðære lyfte is æfre scinende of
ðam heofenlicum tunglum;

---

8. C. Witodlicæ   A. simble, C. æfre *for* symle   C. ðe *deest*
B. he *deest*   ABD. eall   A. eallinga, BCD. eallunga
9. ABCD. Dæghwam-   B. byð   A. wexende   ABCD.
waniende   AB. prican   A. leoman—fram ðære sunnan
desunt
10. BCD. dæghwamlice   A. feala   A. prican   AD. forþam, B. forðam   C. þe *for* ðe seo   B. micle   A. fur-
þor *for* ufor   CD. si
11. A. foran, CD. forn   B. geanunga foron þa sunnan *for* foron
ongean þære sunnan   BD. hyre   C. ontenð   A. byþ,
B. byð
12. A. Simble, C. Symble   A. hrycg, BD. hrigc   C. ðe
B. onlyht   A. byþ, B. byð
13. C. niwe   C. ðe   C. ðe he his   BC. hwyrfe
14. A. fæcc   A. bufan   A. heofonlicum

terrae transgressa, rursus a sole cernatur. Uerum ne hoc omni plenilunio patiatur, latitudinem ei signiferi,[1] quae XII partium est, diuersamque apsidum altitudinem succurrere. *DTR. vii.*

8. Lunam non minui, nec crescere dicunt, sed a sole illustratam, a parte quam habet ad eum, paulatim uel ab eo recedendo, uel ei appropinquando, nobis candidam partem reuoluere, uel atram. *DNR. xx.*

9. Quia enim prima luna, inquiunt, quatuor punctos lucet, adiicitur hic numerus a secunda luna quotidie usque ad plenilunium, detrahiturque dehinc paribus spaciis in diminutionem. *DTR. xxiv.*

10. Luna quotidie quatuor punctis, siue crescens a sole longius abit, seu decrescens soli uicinior, quam pridie fuerat redditur. *DTR. xvii.*

11. *DTR. xvii passim.*

12. Semper enim luna auersis a sole cornibus, rotundam sui partem pandit ad illum. *DTR. xxv.*

13. Sed quando cum illo (*sc.* sole) est, eam partem ad nos habere, qua non illustratur, et ideo nihil in ea lucis uideri. *DTR. xxv.*

14. Circa fines telluris solis splendor undique diffusus, ea libere quae tellure procul absunt aspiciat. Ideoque aetheris quae ultra lunam sunt spatia, diurnae lucis plena semper efficiat: uel suo uidelicet, uel siderum radiata fulgore . . . Purissimum illud et proximum coelo inane, diffusis ubique siderum flammis, semper lucidum fit. *DTR. vii.*

[1] Zodiacus uel signifer. *DNR. xvi.*

9–10. *Cp. Byrhtferð, pp. 160–4, MS. Caligula A. XV, fol. 126b–127a* (*Leechdoms, iii. 222 sq.*), *and The Leofric Missal, ed. Warren, p. 55.* 12. *Cp. viii. 9 below.* 14. Nam supra lunam, quae aeris aetherisque confinio currit, omnia pura ac diurnae lucis sunt plena. *DNR. xxv.*—Superiora tamen illa spatia, quae aetheris nomine censentur, et a turbulento hoc aere usque ad sidereum pertingunt coelum, semper ob siderum circumeuntium reddunt[ur] lucida fulgorem. *Bede (Giles vii. 14).*

15. Hit getimað hwiltidum þonne se mona beyrnð on ðam ylcan strican þe seo sunne yrnð . þæt his trendel underscyt ðære sunnan to ðan swiðe þæt heo eal aðeostrað · 7 steorran æteowiað swylce on nihte;

16. Ðis gelimpð seldon · 7 næfre buton on niwum monan;

17. Be ðam is to understandenne þæt se mona is ormæte brâd · þonne hê mæg ðurh his underscyte þa sunnan aðeostrian;

18. Seo niht hæfð seofon dælas · fram ðære sunnan setlunge oð hire upgâng;

19. An ðæra dæla is Crepusculum · þæt is æfengloma;

20. Oðer is Uesperum · þonne se æfensteorra betwux þære repsunge æteowað;

21. Ðridde is Conticinium · þonne ealle ðing suwiað on heora reste;

22. Feorða is Intempestum · þæt is midniht;

---

15. A. hyt getymaþ   C. þe   A. stricon   C. ðeo   C. is *for* his   A. underscytt   A. þa *for* ðære   C. sunnan t'e'ndel to *for* sunnan to   ABD. þam   A. swyþe   ABD. eall A. aþystraþ   C. 7 þa steorran   AC. ætywaþ   ACD. swilce
16. A. niwne, C. niwe
17. C. understandanne   C. ðe   A. monna   A. ormætlic, C. ormætlice   A. heo   AC. aþystrian
18. C. þeo   BC. seofan   BC. from   B. settlunge ACD. hyre
19. AC. þære   C. dæla *deest*   D. Crepulum
20. C. uesperum · þæt is þonne, D. Vesperum · þæt is æfen; þonne C. se *deest*   D. þære *deest*   A. ætywaþ, C. ætywþ
21. A. þridda   C. þone   B. sweowiað, C. swigiað B. hyra
22. A. feorþe, C. Feorðe

15–16. Solem interuentu lunae, lunamque terrae obiectu nobis perhibent occultari: sed solis defectum non nisi nouissima primaue fieri luna, quod uocant coitum: lunae autem non nisi plena. *DNR. xxii.*

17. Non posse autem totum solem adimi terris intercedente luna, si terra maior esset quam luna. *DNR. xxii.*

18. Cuius (*sc.* noctis) partes sunt septem:

19. Crepusculum, id est, dubia lux inter lucem et tenebras, nam creperum, dubium dicimus.
20. Uesperum, apparente stella huius nominis.

21. Conticinium, quando omnia silent.

22. Intempestum, quod est medium et inactuosum noctis tempus.

---

15. Manifestum est solem interuentu lunae occultari, lunamque terrae obiectu, ac uices reddi, eosdem solis radios luna interpositu suo auferente terrae, terraque lunae. *DTR. xxvii.—Cp. ad iii. 6 above.*
16. Certum est . . . solis defectum, non nisi nouissima, primaue fieri luna. *DTR. xxvii.*
17. Non posset quippe totus sol adimi terris intercedente luna, si terra maior esset quam luna. *DTR. xxvii.*
18–25. *Cp. also DTR. vii in fine.*—Seo niht hafað seofon todæled-nyssa: crepusculum ys seo forme, þæt ys æfengloma; oðer ys uesperum, þæt ys æfen oððe hrepsung; þridde conticínium, þæt ys switíma oððe salnysse tíma; feorðe intempestiuum, þæt ys midniht oððe unworclic tima; fifte gallicinium, þæt ys hancred; þonne sceolon góde munecas arísan 7 Gode singan; syxte matutinum uel aurora, þæt ys dægred; þonne eac gewuniað þa syfre Godes þegnas mid mode 7 stefne God to wurðian 7 Benedictus Dominus bliðelice upahebban; seo seofoðe ys þære nihte todælednyss diluculum geciged, þæt ys ærne mergen betwux dægrede 7 þære sunnan uppgange. *Byrhtferð, pp. 122–4.*—Dies. dæg. Caligo. dimnes. Mane. ærmyrgen. Crepusculum. dægred. *uel* tweone leoht. *uel* þeorcung. Conticinium. hancred. *uel* Gallicantum. Uesperum. *uel* Serum. æfen. Intempestum midniht Gallicinium. hancred. Matutinum uhtentid Diluculum. dægred Aurora. dægrima. *MS. Additional 32246, fol. 18v–19r.*

23. Fifta is Gallicinium · þæt is hâncred;

24. Sixta is Matutinum · uel Aurora · þæt is dægred;

25. Seofoða is Diluculum · þæt is se ærmergen · betwux ðam dægrede · ⁊ sunnan upgange;

26. Wucan ⁊ monðas sind mannum cuðe æfter heora andgite · ⁊ ðeah ðe we hî æfter bôclicum andgite awriton · hit wile ðincan ungelæredum mannum to deoplic ⁊ unge-wunelic;

27. We secgað swa ðeah be ðære halgan eastertide · þæt swa hwær swa bið se mona feowertyne nihta eald · fram duodecima kalendas aprelis · þæt on ðam dæge bið · seo easterlice gemæru · þe we hâtað terminus;

28. ⁊ gif se terminus þæt is se quarta decima luna becymð on ðone sunnandæg · þonne bið se dæg Palmsunnandæg;

29. Gif se terminus bescyt on sumum dæge þære wucan · ðonne bið se sunnandæg þær æfter easterdæg;

---

23. A. gallicicinium   A. hancræd, C. hanecræd
24. BD. Syxta   C. ut *for* uel   A. dægræd
25. CD. se *deest*   A. ærne mergen, B. ærmærien, C. ærne mor-gen, D. ærne merien   B. betweox   C. ðam *deest*   A. dægræde
26. A. syndon, BCD. synd   B. hyra   B. andgyte   A. hig
B. andgyte   A. awritan   D. wyle   A. þincean, B. þingcan
27. G. *a subtitle, clearly referring to §§ 27–29, is added in the margin*: DE PASCHE DOMINI INUENIENDO   C. §§ *27–29 desunt*
A. secgeaþ   A. hwar   B. þe mona byð *for* bið se mona
AD. aprilis   B. byð
28. B. gyf   A. se *deest*   A. becymbþ   A. þonne *for* ðone
A. byþ, B. byð   A. dæg on palm-
29. B. Gyf   A. gescytt, BD. gescyt   A. sunnan *for* sumum, B. sunnon, D. sumon   B. byð

23. Gallicinium, quando gallus resonat.

24. Matutinum, inter abcessum tenebrarum, et aurorae aduentum:

25. et diluculum, quasi iam incipiens parua diei lux, haec et Aurora, solem praecedens. *DT. iii.*

26. *Refers to DTR. viii–xv, and to DT. iv–vi.*

27–29. Terminus quartadecimalis lunae paschae.—Omnis luna quae accensa fuerit ab octaua idus martii usque nonas aprelis, dominicus dies qui sequitur illam quartam decimam lunam dies erit paschae, et si bissextus fuerit uel non.—Post duodecima kalendas aprelis in quale datarum lunam quartam decimam inueneris, ibi fac terminum paschale.—Ubicumque inueneris quartam decimam lunam post .XXX. lunam martii, ibi fac terminum paschę, et in sequenti dominica erit pascha. *Leofric Missal, ed. Warren, p. 48.*

28. Conuenit itaque diligenter aduertere, ut quoties decimaquarta luna in dominicum incurrit diem, in sequentem septimanam paschalem diem potius differamus. *DTR. lix.*

---

23. *Cp. viii. 4 below.*
25. *Cp. vii. 5 below.*
27. *Cp. vi. 2–4 below.*
27–29. Nam quae post XIIII lunam dominica dies occurrit, ipsa est paschalis dominicae resurrectionis dies. Quae quidem decimaquarta luna primum in aequinoctio, id est, duodecimo Calendarum Aprilium . . . suum uespere processum terris ostendit. *DTR. lix.*— *Cp. DTR. xlv.*—He gyme æfter .XII. kl. Aprelis hwær beo se mona feowertyne nihta eald, ⁊ wite eac þæt he byð þæt gemǽre þæs termenes Pasche. Gyf he byð on Sunnandæg luna .XIIII., þænne anbydie we þæt se oðer Sunnandæg cymð; gif he byð on Þunresdæg, þonne do we þæt ylce. *Byrhtferð, p. 136.*—*Cp. Bede, De Paschae Celebratione Liber, Migne, xc. 599–606.*

[IV.] DE ANNO.

1. Þære sunnan gear is þæt heo beyrne ðone micelan circul
Zodiacum · ⁊ gecume under ælc þæra twelf tacna;
2. Ælce monað heo yrnð under an ðæra tacna;
3. An ðæra tacna is gehâten · Aries · þæt is Ramm;
4. Oðer · Taurus · þæt is fearr;
5. Ðridda · Gemini · þæt sind getwisan;
6. Feorða · Cancer · þæt is Crabba;
7. Fifta · Leo;
8. Sixta · Uirgo · þæt is mæden;
9. Seofoða · Libra · þæt is pund · oþþe wæge;
10. Eahteoðe is Scorpius · þæt is ðrowend;
11. Nigoðe is Sagittarius · þæt is Scytta;
12. Teoðe is Capricornus · þæt is buccan horn · oððe bucca;
13. Endlyfte is Aquarius · þæt is wætergyte · oþþe se ðe
wæter gŷt;
14. Twelfte is Pisces · þæt sind fixas;

IV. F begins here   AC. title deest
1. C. þonne   BF. miclan   C. þære, F. þara
2. D. the whole paragraph deest   A. Ælcon monþe   C.
þære, F. þara
3. F. þara   B. ys gehaten   A. ys ramm   CDF. ram
4. C. fear
5. A. þrydda, C. Ðridde, D. þridde, F. þæt þridda   A. syndon, BCDF. synd   ADF. getwysan, C. getwinan
6. C. Feorðe
7. C. Fifta leo · þæt is leo
8. BDF. Syxta
9. F. lybra   A. oþþ
10. A. Eahtoþa, BD. Eahtoðe, C. Eahtoða, F. Eahtaðe   ABCDF.
is deest
11. BCDF. Nigoða   CF. is deest   F. Sagyttarius   C.
svytta · oððe sceotere
12. B. Teoða   B. ys, C. is deest
13. A. Endlifte, B. Endlyfta, C. Endlufte, D. Endleofte   AC.
is deest, F. ys   A. wæterscyte   B. þe þe for se ðe   A. gytt,
F. geot
14. C. Twelte, D. Twelfta   A. is deest   ABDF. synd   AF.
fyxas

1–14. Singuli autem menses sua signa, in quibus solem recipiant, habent: Aprilis, Arietis: Maius, Tauri: Iunius, Geminorum: Iulius, Cancri: Augustus, Leonis: September, Uirginis: October, Librae: Nouember, Scorpionis: December, Sagittarii: Ianuarius, Capricorni: Februarius, Aquarii: Martius, Piscium. *DTR. xvi.*

1–14. *Cp. DNR. xvii.*—An circul ys þe uðwitan hatað zodiacus, oððe horoscopus, oððe Mazaroth, oððe sideralis. Þurh þæne yrnð seo sunne . . . Þes circul ys todæled on twelf, 7 seo sunne geyrnð þas twelf fætu binnan .XII. monðum. *Byrhtferð, p. 4.*—*Cp. ibid., pp. 114–16.*

15. Ðas twelf tacna sind swa gehîwode on ðam heofonlicum rodere · 7 sind swa brade · þæt hî gefyllað twâ tida mid
hire upgange · oððe niðergange;

16. Ælc ðæra twelf tâcna hylt his monað · 7 þonne seo
sunne hî hæfð ealle underurnen · ðonne bið an gear agân;

17. On ðam geare sind getealde twelf monðas · 7 twâ 7

f. 257 v. fiftig wucan · þreo hund daga · 7 fif 7 sixtig daga · | 7 þær to
eacan six tida · þe maciað æfre embe þæt feorðe gear þone
dæg 7 ða niht þe we hatað bissextum;

18. Romanisce leoda onginnað heora gear · æfter hæðenum
gewunan · on winterlicere tide;

19. Ebrei healdað heora geares anginn · on lenctenlicere
emnihte;

20. Þa Greciscan onginnað heora gear æt ðam sunstede ·
7 ða Egyptiscan on hærfeste;

21. Ac ða Ebreiscan ðeoda þe Godes · æ heoldon ongunnon
heora geares anginn ealra rihtlicost · þæt is on ðære lencten-
lican emnihte · XÎI · kalendas aprilis · on ðam dæge þe seo

---

15. A. synt, BCDF. synd.	A. heofenlican, BCD. heofenlicum,
F. heofonlican	B. roderum	A. syndon, BCD. synd	C.
tide	AC. heora, BD. hyra, F. hyre	AB. nyþer-, D. nyðer-
16. F. þara	D. healt, F. hilt	BF. byð
17. A. syndon, BDF. synd	C. monðas · twa for monðas · 7
twa	C. anfiftig for 7 fiftig	A. fifti, F. fyftig	B. wucena
C. daga synd · 7 for daga · 7	BDF. syxtig, C. syxti	BDF syx
B. þa	ABCD. ymbe
18. A. Romonanisce	A. leode, B. leodan	BCD. ongynnað,
F. ongynneð	F. hyra	ACDF. winterlicre
19. F. hebrei	AF. hyre	AC. angin, B. annginn, F. angynn
A. lentenlicre, BD. lenctenlicre, C. lenctelicre, D. læntenlicre emnyhte
20. A. greciscean	CDF. ongynnað	BF. hyra	A. sum-
stede, B. sunnstede, F. stede (sun deest)	A. egiptiscean, B. egip-
tiscan, C. egyptiyscan	B. hærfest
21. B. Ac deest	A. þu for ða	A. ebreiscian	C. ðeoda
deest, F. þeode	B. agunnon	A. heara, F. hyra	A. angin
C. geares anginn desunt, F. gear for geares anginn	A. ristlicost
C. lenctenlican deest	D. emnyhte	AD. aprilis, F. aprelis

15. Tantae sunt magnitudinis, ut non minore quam duarum spatio horarum, uel oriri, uel occidere, uel de loco possint moueri. *DTR. xvi.*

16. Singuli autem menses sua signa, in quibus solem recipiant, habent. *DTR. xvi.*

16–17. Item solis est annus, cum ad eadem loca siderum rediit, peractis CCCLXV diebus, et VI horis, id est, quadrante totius diei, quae pars quater ducta cogit interponi diem unum quod Romani bissextum uocant, ut ad eundem circuitum redeatur. *DTR. xxxvi.*

18–20. Annum autem civilem, id est, solarem, Hebraei ab aequinoctio uerno, a solstitio Graeci, Ægyptii ab autumno, a bruma incipiunt Romani. *DTR. xxxvi.*

16–17. On þam geare þe man hǽt solarem on Lyden beoð þreo hund daga 7 fif 7 syxtig daga, 7 syx tida. Þa synd on Lyden quadrantes genemned. Of þissum syx tidum aspringð up bissextus. *Byrhtferð, p. 64.—Cp. ibid., p. 2.—Cp. vii. 5 below.*

17. Completusque solis annus non CCC solum et LX diebus, sed additis V diebus, et quadrante perficietur. *DTR. xvi.—Cp. Studien zum ae. Computus, pp. 65–8.*

18–20. *Cp. DT. ix.*

18–21. We habbað oft gehyred þæt men hatað þysne dæg (*sc. January 1st*) geares dæg, swylce þes dæg fyrmest sy on geares ymbryne; ac we ne gemetað nane geswutelunge on cristenum bocum, hwî þes dæg to geares anginne geteald sy. Þa ealdan Romani, on hæðenum dagum, ongunnon þæs geares ymbryne on ðysum dæge; and ða Ebreiscan leode on lenctenlicere emnihte; ða Greciscan on sumerlicum sunstede; and þa Egyptiscan ðeoda ongunnon heora geares getel on hærfeste . . . Rihtlicost bið geðuht þæt þæs geares anginn on ðam dæge sy gehæfd, þe se Ælmihtiga Scyppend sunnan, and mônan, and steorran, and ealra tida anginn gesette; þæt is on þam dæge þe þæt Ebreisce folc heora geares getel onginnað. *Homilies, i. 98.—Isti anni (sc.* solaris et lunaris annus) utrique incipiunt a principio pascalis festi ibidemque finiuntur . . . Romanorum circulus, qui dicitur lunaris, Ianuario incipit ibique finitur . . . Romanisce men habbað heora circul, 7 se fehð on Ianuario 7 þær eft geendað. *Byrhtferð, pp. 18, 20, 22.—But cp. also*: Ærest we wyllað fón on Ianuarium, forþon he ys heafodhebba 7 eac þæs geares geendung. *Ibid., p. 62.—Cp. ibid. 20, 19 and 80, 3.*

sunne 7 [se]¹ mona 7 ealle tunglan 7 gearlice tida gesette
wæron;

22. Soðlice þæs monan gear hæfð seofon 7 twentig daga 7
eahta tida;

23. On ðam fyrste he underyrnð ealle ða twelf tacna · þe
seo sunne undergæð twelf monað;

24. Se mona is soðlice be sumon dæle swiftra ðonne seo
sunne · ac swa ðeah þurh ða swyftnysse ne mihte hê underyrnan ealle ða twelf tunglan · binnon seofon 7 twentigum
dagum 7 eahta tidum · gif hê urne swa ûp swa seo sunne;
25. Ðære sunnan ryne is swiðe rûm · forðan ðe heo is swiðe
upp;
26. 7 þæs monan ryne is nearo · forðan þe hê yrnð ealra
tungla nyðemyst · 7 þære eorðan gehendost;

27. Nu miht ðu understandan · þæt læssan ymbgang hæfð se
man þe gæð onbuton an hûs · þonne se ðe ealle þa burh begæð;

¹ G. *deest*

---

21 (*cont*.)   BG. se *deest*   AC. tungla
22. D. tida *for* daga
23. A. *written over erasure* fyrmest *for* fyrste, C. furste
24. A. sunnon *for* sumon, B. suman, F. sumum   ABCF.
swiftre, D. swyftre   F. þeh   ADF. swiftnesse, BC. swiftnysse
F. ne ne mihte   C. underyrnon   AC. tungla   C. eallam þa
twel tungla   ABD. binnan   C. seofen   C. twentig   BF.
gyf   A. upp   A. seo *deest*, B. þeo   A. sunne deþ, BCDF.
sunne deð
25. C. rune, F. rine   A. swyþe   AD. forþam, F. forþon
AC. swa *for* swiðe   BCDF. up
26. C. þes   F. rine   B. is swiðe nearo   AD. forþam
A. heo   A. yrð   A. tungla *deest*   A. nyþemest, B. niðemest,
CD. nyðemest, F. neoþemæst   C. gehændast, F. gehændost
27. A. hu *for* Nu   F. myht   D. ymbegang, F. embegang
F. hæf   ABD. mann   AB. abutan. C. butan   C. ðe *for* se
B. eall   C. begæþ

22–23. Sed lunaris annus quadrifarie accipitur: primus est namque, cum luna XXVII diebus, et VIII horis Zodiacum percurrens, ad id signum ex quo egressa est reuertitur. *DTR. xxxvi.*

23. Nunc dicamus, quia sol CCCLXV diebus et VI horis, luna XXVII diebus, et VIII horis Zodiaci ambitum lustrant. *DTR. xvi.*

24. *DTR. xviii passim.*

25–26. Nec mirari opus est, cum lunam . . . multo inferius ac uicinius terrae quam solem . . . circuire uiderimus: quia uidelicet multo inferius non solum Sole, sed et Uenere ac Mercurio, quae infimae stellarum sunt, luna in confinio aeris huius turbulenti, et puri decurrit aetheris. *DTR. xxvi.*

22, 26, 27. Se seofoþa is se môna, ealra tungla nyþemest, and forðî hæfþ læstne embegang, and forþŷ hê gefylð his ryne on seofon and twentigum dagum and eahta tîdum. *Interrogationes, xxi.*

26. *Cp. iii. 7 above.*

28. Swa eac se mona hæfð his ryne hraðor aurnen on ðam læssan ymbhwyrfte · þonne seo sunne hæbbe on ðam maran ; 29. Þis is þæs monan gear · ac his monað is mare · þæt is ðonne hê gecyrð niwe fram ðære sunnan oð þæt hê eft cume hire forne gean · eald 7 ateorod · 7 eft ðurh hî beo ontend ; 30. On ðam monðe sind getealde nigon 7 twentig daga · 7 twelf tida · þis is se mônlica monað · 7 his gear is þæt hê underyrne ealle ða twelf tunglan ; 31. On sumon geare bið se mona twelf siðon geniwod · fram ðære halgan eastertide · oð eft eastron · 7 on sumon geare hê bið þreottyne siðon geedniwod ; 32. Þæt gear þe we hâtað Communis · hæfð twelf nîwe monan · 7 þæt gear ðe we hâtað embolismus · hæfð þreottyne niwe monan ;

33. Se mônlica monað hæfð æfre on ânum monðe þritig nihta · 7 on oðrum nigon 7 twentig ;

34. On swa hwilcum sunlicum monðe swa swa se mona geendað · se bið his monað ;

---

28. F. rine   B. raðor, F. raþor   F. aurnnen
29. A. hyrne *for* hire, BF. hyre, D. *deest*   CD. forn ongean
A. agean   A. æteorod, D. ateoraðð   A. hig   A. byþ *for* beo
F. ontent
30. ABCDF. synd   D. geteald   C. nigan   ABDF.
monelica   B. hys   AC. tungla
The second fragment of H *begins here*
31. BFH. sumum   A. geare he byþ   BCF. byð   A.
syþon, H. siðum   H. þare   H. eastertit   H. eastran
A. on *deest*   BFH. sumum   A. byþ, F. byð   F. þreottene
F. syðon, H. siðan   B. geedniwad
32. H. gęr   AC. þe *deest*   H. hatat   A. communus ·
þæt þæt hæfð, C. communis þæt hæfð   H. gear *deest*   AC. ðe *deest*
C. we hâtað *desunt*   AC. embolismus · þæt hæfð   F. þreottene
33. ABFH. monelica   A. monoþ   A. þritti, D. þrittig, F.
þriti   CF. nigan   A. twenti
The second fragment of H *ends here*
34. AD. hwylcum   C. sunlicum *deest*, F. sunlycum   ABCDF.
swa *for* swa swa   F. geændað   A. byþ, BF. byð   A. hys
A. monoð

29–30. Secundus (*sc.* lunaris annus) duobus. diebus, et quatuor horis prolixior, qui consuete mensis appellatur, cum solem, a quo noua digressa est, XXIX diebus, et XII horis exactis iam defecta repetit. *DTR. xxxvi.*

31–32. Tertius (*sc.* lunaris annus), qui XII mensibus huiusmodi, id est, diebus CCCLIIII expletur, et uocatur communis, eo quod duo saepissime tales pariter currant. Quartus qui ἐμβόλισμος Graece dicitur, id est, super augmentum, et habet XIII menses, id est, dies CCCLXXXIIII qui uterque apud Hebraeos a principio mensis paschalis incipit, ibidemque finitur. *DTR. xxxvi.*

33. Menses propter lunae circulum, qui XXVIIII semis diebus constat, tricenis undetricenisque diebus alternantes. *DTR. xi.*

---

29. Ideoque rectius ita definiendum, quod mensis lunae sit luminis lunaris circuitus, ac redintegratio de noua ad nouam. *DTR. xi.*

32. Þæt þridde [gér ys] communis, þæt ys gemæne gér. Þonne beoð þi geare þreo hund daga 7 feower 7 fiftig daga fram Eastertíde þæt he eft cume; 7 þonne hyt byð embolismus oððe embolismaris, þæt ys eal án, þonne beoð þi geare þreo hund daga [7 feower 7 hundeahtatig]. *Byrhtferð, p. 20.—Cp. ibid., pp. 78–80.*

33. Ianuarius. Martius. Maius. Iulius. September. Nouember. Recte tenent tricesimam lunam. Februarius. Aprilis. Iunius. Agustus. October. December. Rite pertinent ad uigesimam nonam lunam. *MS. Caligula A. XV, fol. 127v.*—Heræfter we wyllað eow amearkian hwylce monðas habbað þrittig nihta ealdne monan, hwylce nigon 7 twentig. Ianuarius, Martius, Maius, Iulius, September, Nouember—þa sceolon habban þrittig nihta ealdne monan, buton hyt awende se mihtiga embolismus. *Byrhtferð, p. 90.*

34–35. Primum mensem nouorum . . . Nisan appellantes, qui propter multiuagum lunae discursum, nunc in Martium mensem, nunc incidit in Aprilem, nunc aliquot dies Maii mensis occupat. *DTR. xi.*

35. Ic cweðe nu gewislicor · gif se ealda mona geendað twâm dagum oððe ðrim binnon hlydan monðe · þonne bið he geteald to ðam monðe · 7 be his regolum acunnod · 7 swa forð be ðam oðrum; 36. Feower tida sind getealde on ânum geare · þæt sind · UER · ESTAS · AUTUMNUS · HIEMPS; 37. Uer · is lenctentîd · seo hæfð emnihte; 38. Estas is sumor · se hæfð sunstede; 39. Autumnus is hærfest · se hæfð oðre emnihte; |

f. 258 r

40. Hiemps is winter · se hæfð oðerne sunstede; 41. On ðisum feower tidum yrnð seo sunne geond mislice dǣlas · bufon ðisum ymbhwyrfte · 7 þas eorðan getemprað; 42. Soðlice þurh Godes foresceawunge · þæt heo symle on anre stowe ne wunige · 7 mid hire hǣtan middaneardlice wæstmas forbærne; 43. Ac heo gǣð geond stowa · 7 temprað þa eorðlican wæstmas · ægðer ge on wæstme · ge on ripunge; 44. Þonne se dæg lângað þonne gǣð seo sunne norðweard oð þæt heo becymð to ðam tacne þe is gehâten cancer · þær is se sumerlica sunstede · forðan þe heo cyrð þær ongean eft

---

35. A. cwelle *for* cweðe   BF. gyf   F. geændað   ABCDF. binnan   A. byþ, BF. byð, C. bid   D. regole   F. accunnod 36. ABDF. synd   A. synt, BCDF. synd   A. Aestas, B. æstas   BF. hiems, C. hyemps 37. A. ys   C. hæð   DF. emnyhte 38. A. ęstas, B. Ęstas   A. sumstede, B. sunnstede 39. A. *paragraphs 39 and 40 desunt*   B. þe   D. emnyhte 40. B. Hiems, C. Hyemps, F. hiems   B. sunnstede 41. BD. þysum, F. þissum   F. geon   B. mistlice, F. missenlice   A. bufan   BD. þysum, F. þissum   F. embhwyrfte   A. þa 42. AF. symble, C. æfre *for* symle   A. nanre oþre *for* anre stowe   B. gewunige *for* ne wunige   ABDF. hyre   A. hæton   B. middaneardes, CD. middangeardlice   D. wæstmas ne forbærne 43. A. stowe   A. earðlican 44. C. se *for* seo   A. ys   D. þæt *for* þær   A. ys   F. sumorlica   AB. sunnstede   D. forþam, F. forþon   C. hy B. cymð *for* cyrð   C. þer   C. ongean *deest*

36–43. Tempora sunt anni quatuor, quibus sol per diuersa coeli spatia discurrendo subiectum temperat orbem: diuina utique procurante sapientia, ut non semper eisdem commoratus in locis, feruoris ariditate mundanum depopuletur ornatum, sed paulatim per diuersa commigrans, terrenis fructibus nascendis maturandisque temperamenta custodiat. *DTR. xxxv.*

44–46. Sol autem, inquit, ipse quatuor differentias habet: bis aequata nocte diei, uere et autumno in centrum incidens terrae, octauis in partibus Arietis ac Librae: bis permutatis spaciis in auctum diei bruma octaua in parte Capricorni: noctis uero solstitio totidem in partibus Cancri. *DTR. xxx.*

36–40. *Cp. DT. viii.*—Ymbe þa feower timan we wyllað cyðan iungum preostum . . . Ver ys lengtentima . . . 7 he hæfð emniht. Se oðer tima hatte æstas, þæt byð sumor . . . 7 he hæfð sunstede. Se þridda tima ys autumnus on Lyden gecweden, 7 on Englisc hærfest . . . 7 he hæfð emniht . . . Se feorða tima ys genemned hiemps on Lyden 7 winter on Englisc. He hæfð sunstede. *Byrhtferð, pp. 90–2.*

41–43. Hic (*sc.* sol) cursu uariante dies et menses, tempora diuidit et annos, aeris temperiem accedendo uel recedendo pro temporum ratione dispensat: ne si semper in iisdem moraretur locis, alia calor, alia frigus absumeret. *DNR. xix.*

43. On lengtentiman springað oððe greniað wæstmas, 7 on sumera hig weaxað, 7 on hærfest hig rípiað. *Byrthferð, p. 92.*

suðweard · 7 se dæg ðonne sceortað · oð þæt seo sunne cymð · eft suð to ðam winterlicum sunstede · 7 þær ætstent; 45. Ðonne heo norðweard bið · þonne macað heo‚ lenctenlice emnihte · on middeweardum hire ryne; 46. Eft ðonne heo suðweard bið · þonne macað heo hærfestlice emnihte; 47. Swa heo suðor bið swa hit swiðor winterlæcð · 7 gæð se winterlica cyle æfter hire · ac ðonne heo eft gewent ongean · ðonne todræfð heo þone winterlican cyle · mid hire hatum leoman; 48. Se lângigenda dæg is ceald forðan þe seo eorðe bið mid ðam winterlicum cyle þurhgân · 7 bið langsum ær ðan ðe heo eft gebeðod sy; 49. Se sceortigenda dæg hæfð liðran gewederu þonne se langigenda · forðan þe seo eorðe is eal gebeðod · mid þære sumerlican hætan · 7 ne bið eft swa hraðe acoled; 50. Witodlice se winterlica mona gæð norðor · þonne seo sunne gange on sumera · 7 forði hæfð scyrtran sceade þonne seo sunne;

---

44 (*cont.*) A. sceortaþ þonne   A. cympð   F. eft cymð suð   AF.
winterlican   AB. sunnstede   F. ætstænt
   45. A. byþ, BC. byð   F. læntenlice   D. emnyhte   ADF.
middeweardan   ABDF. hyre
   46. C. suðwæard   A. byþ, BCF. byð   A. ærfæstlice, B.
hærfæstlice   D. emnyhte
   47. AF. byð   A. witerlica   ABDF. hyre   C. todrefð,
A. dræfð   A. þæne   ABDF. hyre   D. hatan   F.
leomum
   48. ABCDF. langienda   A. byþ *for* dæg   A. ys   AD.
forþam   C. ðe *for* þe seo   A. byþ, B. byð   AC. winterlican
A. byþ, B. byð   ADF. ær þam, B. ær ðam   B. ef   F. gebeþed
   49. A. sceortgenda, CF. scortigenda, D. sceortienda   A. lyþran,
D. lyðran   A. gewidera, D. gewydera, C. wederu, F. gewideru
A. laˋnˈgienda, BCD. langienda, F. langygenda   ABCDF. *add*
dæg *after* langigenda   AD. forþam, F. forþon   C. ðe *for* þe seo
AF. ys   ABCDF. eall   F. hetan   A. byþ, BF. byð   BF.
raðe   ABCDF. acolod
   50. C. winterlice   C. ðe *for* seo   C. sunnæ   B. ga *for*
gange   C. sumeræ   A. forþi he hæfð, C. forþi he hafað   AD.
sceortran, C. sceortre, F. seortran   C. se *for* seo

50–51. Sed merito quaerere ac mirari sollicitus quisque
potest, quare luna in solstitiali circulo decurrens, tanto altior
aestiuo sole currere, quanto breuiores uideatur facere umbras.
Unde paucis intimandum est, quod hunc lunae progressum
ultra solem in utraque coeli plaga, et australi uidelicet et
septentrionali, Signiferi gignat latitudo, porro in australi
ipsa quoque lunae eiusdem deiectio iuuet . . . Quae (*sc.* luna)
cum australes illius (*sc.* Signiferi) deuenit in locos, aliquanto
humilior hyberno sole apparet. *DTR. xxvi.*

47–48. Winter byð cealdost, lencten hrīmigost, hē byð lengest
ceald. *Cotton Gnomes, ll. 5–6 (B.C. Williams, Gnomic Poetry in Anglo-
Saxon, New York, 1914, p. 127).*

51. Eft on langiendum dagum hê ofergæð ðone suðran sunstede · 7 forði bið nyðor gesewen ðonne seo sunne on wintra;

52. Swa þeah ne gǽð heora naðor ænne prican · ofer ðam þe him geset is · ne dagas ne sind nu naðor ne lengran ne scyrtran · ðonne hî æt fruman wæron;

53. On Egypta lande ne cymð næfre nan winter · ne rênscuras · ac on middan urum wintra beoð heora feldas mid wyrtum blowende · 7 heora orcyrdas mid æpplum afyllede;

54. Æfter heora geripe gæð seo êâ upp Nilus · 7 oferflêt eal þæt Egyptisce land · 7 stent oferflede hwilon monað hwilon leng · 7 siððan to twelf monðum ne cymð þær nan oðer scur · oð þæt seo êâ eft up abrece · swa swa hire gewuna is · ælce geare æne · 7 hî habbað þurh þæt · cornes swa fela swa hî mæst reccað;

[V.] DE MUNDO.

I. Middangeard is gehâten eal þæt binnon þam firmamentum is;

---

51. A. langiende dagan   ABF. heo   A. forgæþ, C. ofergeð
A. þæne   AB. sunnstede   D. forþy   A. forþi he byþ, C.
forþi he byð   B. byð   A. nyþror, C. norþur, D. neoðor, F.
niðor   F. wintre
52. F. þeh   F. hyra   B. naðer   AC. pricon   A. ofor
A. heom   ABD. gesett   A. ys   A. syndon, BCDF. synd
B. naðor *deest*   C. ne *deest*   B. længran   A. þænne   A.
hig   B. wæran
53. AB. egipta   D. næfre *deest*   C. renscyras   A. on
urum middan   ACDF. wintre   ABF. hyra   B. weortum,
F. wirtum   BF. hyra   B. orcerdas, C. orcyerdas, F. orcirdas
C. æwplum
54. F. hyra   A. gieþ   ABCDF. up   A. nilis   A.
oferflett, B. oferfled, D. oferflewð   ABDF. eall   AB. egiptisce
A. stend, F. stænt   F. hwilum   A. monoþ   BC. hwilum
A. lengc, F. læ'n'g   A. seþpan, BD. syððan   C. cumð   A.
upp   A. abrecæ   ABDF. hyre   A. wuna *for* gewuna
A. ys   C. þur   A. feala, F. mycel *for* fela   A. hig   A.
recceaþ, BF. recceað

V. AC. *title deest*

I. ABCF. Middaneard   A. ys   ABDF. eall   F. þæt
deest   ABCDF. binnan   A. ys

53. Sed et Ægyptus nostra hyeme media, maxime campos
herbis floreos, et syluas fertur habere pomis onustas. *DTR.*
xxxv.

53-54. Nilo flumine, quod inter ortum solis et Austrum
enascitur, pro pluuiis utitur Ægyptus, propter solis calorem
imbres et nubila respuens. Mense enim Maio, dum ostia eius,
in quibus in mare influit, Zephyro flante, undis eiectis
arenarum cumulo praestruuntur, paulatim intumescens, ac
retro propulsus, plana irrigat Ægypti: uento autem cessante,
ruptisque arenarum cumulis, suo redditur alueo. *DNR. xliii.*

1, 3. Mundus est uniuersitas omnis, quae constat ex coelo
et terra, quatuor elementis in speciem orbis absoluti globata.
DNR. iii.

1, 3. Mundus ys gehaten eall þæt ys betweox heofenum 7 eorðan 7
on þære sæ. *Byrhtferð, p. 122.*

2. Firmam*entum* is ðeos roderlice heofen mid manegu*m* steorru*m* amet;

3. Seo heofen · 7 sæ · 7 eorðe · sind gehâtene middangeard;

4. Seo firmam*entum* tyrnð symle onbutan us under ðyssere eorðan · 7 bufon · ac þær is ungerim fæc · betwux hire 7 ðære eorðan;

f. 258 *v*. 5. Feower 7 twentig tida | beoð agane · *þæt* is an dæg · 7 an niht · ær ðan ðe heo beo æne ymbtyrnd · 7 ealle ða steorran þe hire on fæste sind turniað onbutan mid hire;

6. Seo eorðe stent on ælemiddan ðurh Godes mihte swa gefæstnod · *þæt* heo næfre ne bihð · ufor ne neoðor · þon*ne* se ælmihtiga scyppend · ðe ealle ðing hylt buton geswince · hî gestaðelode;

7. Ælc sæ̂ þeah heo deop sy · hæfð grund on ðære eorðan · 7 seo eorðe aberð ealle sæ̂ · 7 ðone micclan garsecg · 7 ealle wylspringas · 7 eân ðurh hire yrnað;

8. Swa swa æ̂ddran licgað on þæs mannes lichaman · swa licgað ða wæt*er*æddran geond þas eorðan;

9. Næfð naðor ne sæ̂ · ne eâ · nænne stede buton on eorðan:—

---

2. F. heofon   A. amett
3. F. heofon   ABCF. synd   ABCF. middaneard
4. A. Se   A. symble, C. æfre *for* symle   A. abutan, F. unbuton   A. þisse, CDF. þissere   AB. bufan   A. ys   B. betweox, C. betux, F. betwyx   ABF. hyre
5. A. æfre *for* agane   AC. ær ðan ðe—on fæste *desunt*   D. ær þam   B. æne *deest*   BDF. hyre   A. syndon, BDF. synd, C. sind *deest*   C. þa turniað   A. turniende *for* turniað   A. abutan, F. onbuton   ABDF. hyre
6. D. oðer *for* eorðe   C. eallemiddan   F. myhta   BD. byhð, F. byhþ, A. byþ, C. bið   A. naþor ne ufror ne nyþror, B. ne ufor ne nyðor   C. neoðer   F. ælmyhtiga   C. butan B. swince   F. gestaþolode
7. F. þeh   ACD. þeah þe heo   F. eorþon   A. abyrþ, BDF. abyrð   B. ælce *for* ealle   A. sæs   A. þæne   BF. miclan, D. micclan *deest*   F. garsecg   ABD. wyllspringas, F. wilspringas   AC. ea, F. eann   A. an hig þurhyrnað *for* ðurh hire yrnað   BDF. hyre, C. hi *for* hire
8. A. licgeaþ, B. licgeað   A. licgeaþ   B. þas *for* ða   B. urh *for* geond
⊢9. B. naðer   C. butan

4–5. Coelum . . . a centro terrae aequis spatiis undique collectum. Unde et conuexum mediumque quacunque cernatur, inenarrabili celeritate quotidie circumagi sapientes mundi dixerunt . . . argumento siderum nitentes, quae fixo semper cursu circumuolant. *DNR. v.*

6. Atque ipsa terra, quae mundi media atque ima, librata uolubili circa eam uniuersitate pendet immobilis. *DNR. iii.* —Quia in manu eius sunt omnes fines terrae. *Psalmus xcv. 4.*

7–9. Si quis uero quaesierit ubi congregatae sint aquae,

1–5. Mundus est is qui constat ex coelo et terra, mari cunctisque sideribus. Qui idcirco mundus est appellatus, quia semper in motu est; nulla enim requies eius elementis concessa est. *Isidore, Etym.* III, *xxix.*
2. *Cp. i. 6 above.*
4. *Cp. i. 5 above.*
4–5. Firmamentum . . . næfre ne stent stille on ánum, 7 on anre wendinge ða hwile he æne betyrnð gað witodlice forð feowor 7 twentig tída, ðæt is ðonne ealles án dæg 7 an niht. *Hexameron, 145–8.*
5. *Cp. ix. 3 below.*
6. *Cp. xi. 10–12 below.*—Æqualiter enim (sphaera coeli) ex omni parte fertur esse collecta, et omnia similiter respiciens, atque a centro terrae aequis spatiis distincta; ipsaque sui aequalitate ita stabilis, ut eam in nullam partem declinare undique aequalitas collecta permittat, ac nullo fulcimento subuecta sustentetur. *Isidore, DNR. xii. 4.*—Heo (*sc.* seo eorðe) ne lið on nanum ðinge ac on lofte heo stynt ðurh ðæs anes mihte ðe ealle ðing gesceop. 7 he ealle ðing gehylt butan geswince. *Hexameron, 171–3.*—He (*sc.* God Ælmihtig) hylt mid his mihte heofonas and eorðan, and ealle gesceafta butan geswince. *Homilies, i. 8.*—He is butan hêfe, forðon þe he hylt ealle gesceafta butan geswince. *Ibid., i. 286.*—Quia in manu eius sunt omnes fines terrae. Forðanðe on his handa syndon ealle ðære eorðan gemæru. *Hexameron, 179 sq.*
7–9. Ða sæ he gelogode swa swa heo lið git wiðinnan ða eorðan on hyre ymbhwyrfte 7 ðeah ðe heo brad sy and gebyged gehú 7 wundorlice deop, heo wunað eall swa ðeah on ðære eorðan bosme binnan hyre gemærum. *Hexameron, 181–5. Cp. DNR. xliv.*

[VI.] DE EQUINOCTIIS.

1. Manegra manna cwyddung is · þæt seo lenctenlice
emniht gebyrige rihtlice on octaua kalendas aprilis · þæt is
on Marian mæssedæg · Ac ealle ða Easternan · 7 Egyptiscan
þe selost cunnon on gerimcræfte · tealdon þæt seo lenctenlice
emniht is gewislice on XII̅ᵐᵃ · kalendas aprilis · þæt is on
Sancte Benedictes mæssedæg;

2. Eft is beboden on þam regole · þe us gewissað be þære
halgan eastertide · þæt næfre ne sy se halga easterdæg gemær-
sod · ær ðan ðe seo lenctenlice emniht sy agân · 7 ðæs dæges
lencge oferstige þa niht;
3. Wite nu forði gif hit wære rihtlice emniht on Marian
mæssedæg · þæt se dæg ne gelumpe næfre ofer ðam easter-
dæge · swa swa hê foroft deð;

VI. *The first fragment of* E *begins here* ACE. *title deest* B. DE
AEQUINOCTIIS
1. C. cwiddung   A. ys   F. læntenlice   B. ymniht   C.
gebyrie, F. gebirie   F. aprelis   A. on sancta marian, E. on sancta
maria   ABF. -dæge, E. -deg   A. 7 þa egiptiscean, C. 7 þa egyptiscan, B. egiptiscan, F. egyptiscean   C. tieldon   A. lentenlice
AF. ys   C. gewislice—is *desunt*   F. aprelis   A. ys   A. sanctus
AF. benedictus   B. -dæge
2. C. *paragraphs* 2–6 *desunt* A. ys   A. regule, E. reogole   E.
easterdeg   ADEF. ær þam   F. emnyhte   E. deges   B.
lenge, F. længe   A. oforstige   F. nyht
3. F. Wite þu nu   B. forðy   A. gig *for* gif, BEF. gyf   A.
on sancta marian   ABF. -dæge, E. -dege   E. deg   A. ufor
A. swa *for* swa swa   A. he *deest*   F. forofte

quae omnes terrae partes ad coelum usque cooperuerant;
sciat fieri potuisse, ut terra ipsa longe lateque iussu Creatoris
subsidens, alias partes praeberet concauas, quibus con-
fluentes aquae reciperentur . . . Tellus omnis per inuisibiles
uenas aquis est repleta manantibus. *Bede, Comm. in Lib.
Gen. (Giles vii. 12)*.

1. De aequinoctiis, quod octauo Calendarum Aprilium, et
octauo Calendarum Octobrium: et de solstitiis, quod octauo
Calendarum Iuliarum, et octauo Calendarum Ianuariarum
die sint notanda, multorum late et sapientium seculi, et
Christianorum sententiâ claret . . . Uerum quia sic ut in
ratione paschali[1] didicimus aequinoctium uernale duodecimo Calendarum Aprilium die cunctorum Orientalium
sententiis, et maxime Ægyptiorum, quos calculandi esse
peritissimos constat, specialiter adnotatur. *DTR. xxx.*

2–3. Item catholicae institutionis regula praecipit, ut ante
uernalis aequinoctii transgressum Pascha non celebretur.
Qui igitur VIII Calendarum Aprilium die putat aequinoctium, necesse est idem aut ante aequinoctium Pascha celebrari licitum dicat, aut ante octauum Calendarum Aprilium
diem Pascha celebrari licitum neget. Ipsum quoque Pascha
. . . aut IX Calendarum Aprilium die non fuisse, aut ante
aequinoctium fuisse confirmet. *DTR. xxx.*

[1] Refers to Bede's *De Paschae Celebratione Liber* (Migne, *Patrologia
Latina*, xc). Cp. especially columns 602–5.

1–6. Se ælmihtiga Scyppend gesceop ða tunglan . . . on lenctenlicere emnihte swa swa lareowas secgað, on gerimcræfte XII kł.
Aprilis, and ne beoð næfre eastron ær se dæg cume ðæt ðæt leoht
hæbbe ða ðeostru oferswiðed, ðæt is ðæt se dæg beo lengra ðonne seo
niht. Be ðam oðrum tídum cwyð ðeos ylce bóc . . . *Hexameron, 217–24*.

1–20. *Cp. DT. vii.*
2–4. *Cp. iii. 27–9 above.*

4. Us is neod þæt we ða halgan eas*ter*tide be ðam soðum regole healdon · næfre ær emnihte · 7 oferswiðdum þeostrum;

5. Forði we secgað soðlice þæt seo emniht is swa swa we ær cwædon · on XÍI · *kalendas* a*pri*lis · swa swa þa geleaffullan fæderas gesetton · 7 eac gewisse dægmæl ûs swa tæcað;

6. Eac ða oðre þreo tîda þæt is se sumerlica sunstede · 7 se wint*er*lica · 7 seo hærfestlice emniht sind to emnettenne be ðyssere emnihte · þæt hî syn sume dagas gehealdene ær ðam octaua kalendas;

7. Witodlice se emnihtes dæg · is eallu*m* middanearde ân · 7 gelice lâng · 7 ealle oðre dagas on twelf monðum habbað mislice langsu*m*nysse;

8. On sumu*m* earde hi beoð lengran · on sumu*m* scyrtran for ðære eorðan sceadewunge · 7 ðære sunnan ymbgan*g*e;

9. Seo eorðe stent on gelicnysse anre pinnhnyte · 7 seo sunne glit onbutan be Godes gesetnysse · 7 on þone ende þe heo scinð is dæg · ðurh hire lihtinge · 7 se ende þe heo forlæt bið mid ðeostru*m* oferðeaht · oð þæt heo eft þider genealæce;

---

4. F. ða *deest*   F. halgyan   AEF. soþan, BD. soðan   AEF. healdan   A. oforswiðdum, B. oferswiðum   A. þysru*m*
5. A. secgeað, E. scewiað   F. emnyht   A. ys   A. swa *for* swa swa   F. geleafullan   A. fæderas hyt gesetton   B. dægmælas *for* dægmæl us   E. degmæl   A. tæceaþ, F. tæceað   *The first fragment of* E *ends here*
6. AB. sunnstede   A. hærfæstlice   A. emniht emniht   B. ymniht, F. emnyht   AB. synt, D. synd   A. emnettende ABF. þissere   F. emnyhte   A. hig   A. gehealdenne   B. ær þan   B. actaua
7. F. emnyhtes   AF. ys   D. middangearde   F. angelice *for* an 7 gelice   F. mystlice   A. langnysse, BF. langnisse, CD. langnesse
8. ACDF. sumon   D. gearde   A. hi *deest*   C. byð   B. længran   F. lengran 7 on   ABCDF. sumon   A. sceadwunge, C. sceadunge   A. sunnon   A. ymbegange
9. F. stænt   D. gelicnesse, F. gelycnesse   C. pinhnyte   A. sé   F. glyt   AB. abutan, DF. onbuton   B. abutan gewislice be   AF. gesetnesse   F. ænde   F. scynð ABF. ys   ABDF. hyre   B. lyhtinge, F. lyhttinge   C. þe *for* se F. ænde   A. byþ, B. byð   ABC. þystrum   A. oforþeht, C. oferþeht   F. ẹæft   ACD. þyder, B. ðyder

4-6. Unde nos necesse est ob conseruandam ueritatis regulam, dicamus aperte, et Pascha ante aequinoctium tenebrasque deuictas non immolandum, et hoc aequinoctium duodecimo Calendarum Aprilium diei ueraciter adscribendum, sicut non solum auctoritate paterna, sed et horologica consideratione docemur: sed et caetera tria temporum huiusmodi confinia simili ratione aliquot diebus ante octauum Calendarum sequentium esse notanda. *DTR. xxx.*

7-8. Et quidem aequinoctialis dies omni mundo aequalis et una est: uerum solstitialis et caeteri omnes diuersae longitudinis pro ratione climatum disparium sunt et umbrarum. *DTR. xxxi.*

9. Solertissimus naturalium inquisitor Plinius Secundus qui non negat terram, etsi sit figurae pineae nucis, nihilominus undique incoli . . . *DTR. xxxiv.*

9-10. Causa autem inaequalitatis eorundem dierum terrae rotunditas est: neque enim frustra et in scripturae diuinae, et in communium literarum paginis orbis terrae uocatur. Est enim reuera orbis idem in medio totius mundi positus, non in latitudinis solum gyro, quasi instar scuti rotundus, sed instar potius pilae undique uersum aequali rotunditate persimilis: . . . Talis ergo schematis terra mortalibus ad inhabitandum data, solis circuitus in hoc mundo lucentis certa ratione constitutionis Dei, alibi diem exhibet, alibi noctem relinquit. *DTR. xxxii.*

---

4. Et quae nobis aeternae beatudinis lumen promittit, tunc maxime celebretur, cum solis lumen annuo proficiens incremento, primam sumit de noctis umbra uictoriam. *DTR. lxiv.*

6. Caeteros quoque tres temporum articulos putamus aliquanto priusquam uulgaria scripta continent esse notandos. *DTR. xxx.*

7. Þes monð (*sc.* September) hæfð twelf tida on þære nihte þe seo emniht byð, 7 twelf on dæg. Syððan langað seo niht 7 wanað se dæg eall þæt .XII.ᵐᵃ kl. Ianuarii cymð to mancynne. *Byrhtferð, p. 88.*

9-10. *Cp. DNR. xlvi, and vi. 20 below.*

10. Nu is þære eorðan sinewealtnys · 7 ðære sunnan ymbgang · hremming þæt se dæg ne bið on ælcum earde gelice lang;

11. On India lande wendað heora sceada on sumera suðweard · 7 on wintra norðweard;

12. Eft on Alexandria · gæð seo sunne upprihte on ðam |

f. 259 r. sumerlicum sunstede on middæge · 7 ne bið nân sceadu on nane healfe;

13. Ðis ylce getimað eac on sumum oðrum stowum;

14. Meroe hatte ân igland · þæt is ðæra Silhearwena eard · on ðam iglande hæfð se lengsta dæg on geare twelf tida · 7 lytle mare ðonne ane hea'l'fe tide;

15. On ðam earde ðe is gehâten Alexandria hæfð se lengsta dæg · feowertyne tida;

16. On Italia · þæt is Romana rice · hæfð se [lengsta]¹ dæg fiftyne tida;

17. On Engla lande hæfð se lengsta dæg seofontyne tida;

¹ G. *deest*

10. A. ys   A. sinewealnesse, CF. sinewealtnes   A. ymbegang, B. ymgang   ABF. byð   D. gearde
11. F. hyra   B. scada   A. suþwear
12. ABCDF. up-   BF. sumerlican   AB. sunnstede   A. middan dæge   A. byþ, B. byð   A. nanre, C. anre *for* nane
13. C. *the whole paragraph deest*   F. ilce   D. sumon
14. A. Meloe, C. Mede, F. Merode   A. ys   F. þara   B. sillhearewena   A. land *for* eard, D. geard   C. landæ *for* iglande   A. lengesta, BF. længsta   CF. litle   C. an
15. AC *anticipate the beginning of* § *18*: On þam ylcan earde norþeweardan (C. norðeweardan) *for* On ðam earde ðe is gehaten   D. gearde   C. alexandrie   C. hæf   BF. længsta   C. deg   A. feowertyna, C. feowertune, D. fiftyne *for* feowertyne   C. tide
16. ABC. *the whole paragraph deest*   F. hæfð længysta dæg   G. hæfð se dæg
17. B. *the whole paragraph deest*   F. længesta   A. seofentyne, C. feowertune *for* seofontyne

11. In Indiae gente Oretum, mons est Malaeus nomine, iuxta quem umbrae aestate in austrum hyeme ad septemtrionem iaciuntur. *DTR. xxxi.*

12-13. Simili modo tradunt in Syene oppido quod est supra Alexandriam V millibus stadiorum solstitii die medio nullam umbram iaci, puteumque eius experimenti gratia factum, totum illuminari, ex quo apparet tum solem illi loco supra uerticem esse: quod et in India supra flumen Hypasim fieri tempore eodem Onesicritus scripsit: constatque in Berenice urbe Troglodytarum ... hoc idem ... fieri. *DTR. xxxi.*

14. Rursus in Meroe insula, quae caput gentis Aethiopum ... *DTR. xxxi.*

14-20. Sic fit, ut uario lucis incremento in Meroe longissimus dies XII horas aequinoctiales, et VIII partes unius horae colligat, Alexandriae uero XIIII horas: in Italia XV, in Britannia XVII, ubi aestate lucidae noctes haud dubie repromittunt, id quod cogit ratio credi, solstitii diebus accedente sole propius uerticem mundi, angusto lucis ambitu subiecta terrae continuos dies habere senis mensibus, noctesque e diuerso ad brumam remoto. Quod fieri in Insula Thule Pythias Massiliensis scribit, VI dierum nauigatione in septentrionem a Britannia distante. *DTR. xxxi.*

18. On ðam ylcan earde norðeweardan beoð leohte nihta on sumera · swilce hit ealle niht dagige · swa swa we sylfe foroft gesawon;

19. Thile atte ân igland benorðan þisum iglande · six daga fær on sǣ · on ðam ne bið nân niht · on sumerlicum sunstede six dagum · forðan ðe seo sunne bið þonne swa feorr norð agân · þæt heo hwonlice undergæð · þære eorðan geendunge swilce hit æfnige · 7 þærrihte eft upgæð;

20. Eft on winterlicum sunstede · ne bið nan dæg on ðam foresædan iglande · forðan ðe seo sunne bið þonne swa feorr süð agân · þæt hire leoman ne magon to ðam lande geræcan · for ðære eorðan sinewealtnysse;

21. Is ðeah to witenne þæt symle bið under dæge 7 nihte · feower 7 twentig tida · 7 on emnihtes dæge þæt is ðonne se dæg 7 seo niht gelice lange beoð · þonne hæfð hyra ægðer · twelf tida · swa swa Crist sylf on his godspelle cwæð;

22. Nonne XII. horę sunt diei? La hû ne hæfð se dæg twelf tida?

23. Soðlice þære sunnan ormætan hætu · wyrcð fif dælas on middanearde · þa we hata ð on leden quinque zonas · þæt sind fif gyrdlas;

---

18. AC. ylcan *deest*   D. gearde   A. norþeweardon   C. byð
A. swille, B. swylce   ABC. dagie, F. dagye   C. swa we hit oft
for swa swa we sylfe foroft   F. silue   A. sawon, F. gesawonn
19. C. Tile   ABCDF. hatte   BD. þysum, F. þissum
BDF. syx   A. ofor *for* on   F. sumerlican   AB. sunnstede
BDF. syx   ADF. forþam, B. forðam   A. byþ, B. byð   A.
þænne   CDF. feor   AB. swylce   D. eft *deest*   A. upp-
20. AB. sunnstede   AB. byð   ADF. forþam   AB. byð
C. feor   ABF. hyre   C. mæg *for* magon   C. þan
21. A. ys, B. Ys   F. þeh   C. witanne   A. byþ simble
C. æfre *for* symle   B. byð, F. bið *deest*   ACDF. dæg   B.
'dæg' *added by a later hand. Continental g!*   ABCDF. niht   B.
dæg   A. ys   A. þænne   F. gelyce   C. þonne—cwæð
desunt   D. heora   F. sylua
22. C. *the whole paragraph deest*   ABD. hore, F. hores
23. B. armætan   F. wyrhþ   A. fix *for* fif   D. middangearde   A. þe   C. on leden *desunt*   A. synt, BCDF. synd   F. fyf

21. Sed et noctis utroque tempore quae sit mensura, aeque clarum reddidit, quia necesse est cuiuscunque sint longitudinis dies, simul et nox XXIIII horarum spacium compleant. *DTR. xxxi.*

21–22. Et diem quidem uulgarem Dominus sententia quam et supra posui, definiuit, dicens: Nonne duodecim horae sunt diei? *DTR. v.—Iohannes xi. 9.*

23–26. Quinque circulis mundus diuiditur, quorum distinctionibus quaedam partes temperie sua incoluntur, quaedam immanitate frigoris aut caloris inhabitabiles existunt. Primus est septentrionalis, frigore inhabitabilis, cuius sidera nobis nunquam occidunt. Secundus solstitialis, a parte signiferi excelsissima, nobis ad septentrionalem plagam uersus, temperatus, habitabilis. Tertius aequinoctialis, medio ambitu signiferi orbis incedens, torridus, inhabitabilis. Quartus brumalis, a parte humillima signiferi ad austrinum polum versus, temperatus, habitabilis. Quintus australis, circa uerticem austrinum, qui terra tegitur, frigore inhabitabilis. Tres autem medii circuli inaequalitates temporum distinguunt,

---

21–22. Hora duodecima pars diei est: siquidem XII horae diem complent, Domino attestante, qui ait: Nonne duodecim horae sunt diei? *DTR. iii.*
23–26. *Cp. DTR. xxxiv.*

24. An ðæra dæla is on ælemiddan weallende · 7 unwuni-
gendlic · for ðære sunnan neawiste · on ðam ne eardað nân
eorðlic man for ðam unaberendlicuṃ bryne · þonne beoð on
twa healfa þære hǣtan · twegen dælas gemetegode · naðor ne
to hâte · ne to cealde;
25. On ðam norðran dæle wunað eal mancynn under þam
bradan circule þe is gehaten Zodiacus;
26. Beoð þonne gyt twegen dælas on twa healfa · þam
gemetegodum dælum on suð'e'weardan · 7 norðeweardan þises
ymbhwyrftes cealde 7 unwunigendlice · forðan ðe seo sunne
ne cymð him næfre to · ac ætstent on ægðre healfe æt ðam
sunstedum;

[VII.] DE BISSEXTO.

1. Sume preostas secgað *þæt* bissextus come ðurh *þæt* ·
þæt Iosue abæd æt Gode *þæt* seo sunne stôd stille anes dæges
lencge · ða, ða hê þa hæðenan of ðan earde adylegode · þe him
God forgeaf;

---

24. A. þære, F. þare   AC. dæla *deest*   A. ys   C. eallemiddan   ACF. unwuniendlic, B. ungewunelic   A. on *for* for
AB. neaweste   ABD. mann   A. unaberiendan, B. unberend-
licum, C. unaberendlican   C. brune, F. brine   C. Ðone   F.
cole *for* cealde
25. ABCDF. eall   ABF. manncynn, C. mancun, D. manncyn
A. gehaten *deest*
26. A. þænne   C. healfe þam   B. dæle *for* dælum   A. suþ-
weardan   A. 7 on norþweardan   AB. þyses, F. þisses   C.
mbhwirftes, F. embhwyrftes   A. cealde cealde   ABCF.
unwuniendlice   ADF. forþam   D. ðe *deest*   C. sy *for* seo
A. on ægðre healfe *desunt*   C. æþre heafe   A. sunnstede, C.
sunstede, B. sunnstedum

VII. AC. *title deest*
1. A. secgeaþ   A. iosuę   F. abæde   AF. lenge, C.
længe   A. on *for* of   ACDF. þam, B. ðam   D. gearde
A. adyllogode, BF. adilegode, C. adilogode   A. swa *for* þe

cum sol hunc solstitio, illum aequinoctio, tertium bruma
teneat. Extremi enim semper sole carent. *DNR. ix.*

1–8. Bissextum non ob illum diem fieri ut quidam putant
cum iosuae solem orabat stare credendum est. quia ille dies
fuit et preteriit. sed ab hoc dicitur bissextus. quod in uno-
quoque mense punctus unus crescit. Punctus uero unus
quarta pars horae est. IIII^{or} uero puncti unam horam
faciunt. XII uero puncti III horas explicant. Ergo in IIII^{or}
annis IIII^{or} terne horae quod sunt XII. Diem faciunt
unum qui additur februario. Cumque VI kal. mart. habuerit
et ut in crastino sic habeat uerbi gratia si hodie VI kal. mart.

1–2. Iosue se heretoga mid sige wearð gebyld, and cwæð to ðære
sunnan mid swiðlicum worde, þæt heo of ðære stowe styrian ne
sceolde, ǽrðan ðe his fynd feallende swulton. Þa stôd seo sunne
swiðe healic ongean Gabaô, be Godes hǽse, anes dæges fæc, ǽrðan ðe
heo yrnan dorste, oðþæt ða sigefæstan heora fynd aledon. *Homilies,*
ii. 214.—On ðam dæge bæd Iosue his Drihten, Ᵹ ðus cwæð: Ne astyra
ðu sunne of ðam stede furðor ongean Gabaon, Ᵹ ne gang ðu mona
ongean Achialon ænne stæpe furðor! Ða stod seo sunne on ðam stede
fæste Ᵹ se mona gelice, oð ðæt hi aledon heora fynd. Næs swa lang
dæg ær ðam on ðisum life æfre, ne syððan on ðisre worulde, for ðan
ðe God wolde ða fylstan his cempan Ᵹ feohtan for Israhel. *Joshua x.*
12–14 (ed. Crawford, p. 393).

2. Soð þæt is þæt seo sunne ða stôd anes dæges lencge bufon ðære byrig Gabaon · ðurh þæs ðegenes bene · ac se dæg eode

f. 259 v. forð swa swa oðre dagas · | 7 nis næfre þurh þæt bissextus · þeah ðe þa ûngelæredan swa wenon;

3. Bis · is tuwa; Sextus · se sixta; Bissextus tuwa six · forðan ðe we cweðað on þam geare nu todæg .VI. kalendas martii · 7 eft on merigen .VI. kalendas martii · forðan þe æfre bið ân dæg 7 ân niht mâ on ðam feorðan geare · þonne wære on ðam ðrim ær;

4. Se dæg 7 seo niht weaxað of þam six tidum þe ælce geare beoð to lafe to eacan þam ðrim hund dagum · 7 fif 7 sixtig dagum;

5. Seo sunne beyrnð þa twelf tâcna · on ðrim hund dagum · 7 fif 7 sixtig dagum · 7 on six tidum · swylce heo nu togeare gange on ærne merigen on þæs emnihtes circule · oðre geare

---

2. A. þæt deest   A. ys   C. sy for seo   AF. ða deest
AC. stod stille anes   B. læncge, F. læn'c'ge, C. lenge   A.
bufan   A. gabao   AC. þegnes   F. bede for bene   A.
nys   F. þæt deest   C. þeah—wenon desunt   F. þeh   F.
ðe deest   A. wenaþ, BDF. wenað
The second fragment of E begins here
3. A. ys   A. twia, B. twuwa, CE. twa, DF. tua   E. sextas
ABDEF. syxta   C. is six for se sixta   A. twia, B. twuwa,
DF. tua, E. twa   C. is twa syx for tuwa six   B. syx
ABDF. forþam   B. ðe deest   A. cwiþaþ   E. todeg
DE. 7 deest   A. 7—martii desunt   B. amerigen for on merigen,
F. amergen, C. tomorgen   ADE. forþam, F. forþon   A. biþ
æfre   B. byð   C. daga 7 nihta for an dæg 7 an niht   E. deg
A. þænne
4. E. deg   C. þy for seo   F. nyht   A. wexað   AE.
on for of   ABD. syx   C. byð   A. to lafe desunt   C. þrym
BF. syxtig, AE. sixtigum, D. syxtigum   B. daga
5. A. Seo—sixtig dagum desunt   E. Se for Seo   F. tacnu
F. fyf   B. syxtig, D. syxtigum, E. sixtigum   B. daga   D. syx
CDEF. swilce   C. hy   AF. mergen, BD. merien, C. morgen
F. oð for on   F. þes   B. cyrcule   A. circule · 7 oðre

additur ille dies qui quarto anno expletus est. ut et in crastino sic habeat VI kal. mart. et ideo bissextus dicitur quia bis legitur VI kal. mart. . . . Quid ergo efficit bissextum. hoc est tarditas cursus solis. Inde augustinus dicit. breui autem et uulgari ratione bissextum retardatio generat solis. non ad eandem lineam caeli per CCCLXV dies plene redeuntis a qua recurrit. *Missal of Robert of Jumièges, ed. Wilson, p. 42.*

3–8. Quadrantis autem huiusce per quadriennium in diem integrum colligendi, et in loco suo intercalandi haec est ratio, quia sol annuum coeli ambitum, id est, XII signa circuli Zodiaci notissima, non in CCCLXV diebus, sed superadditis VI horis adimplere cognoscitur: unde fit, ut si uerbi gratia, nunc aequinoctialem coeli locum mane oriens intrauèrit, in hunc anno sequente meridie, tertio uespere, quarto media nocte, quinto rursus in exortu diei, utpote completo diei totius circuitu recurrat, sique necessario diem superfluum admoneat ubilibet interponendum, annique quarti plenitudini esse copulandum. Quem . . . Romani autem VI Calendarum Martiarum die, unde et bissextum uocant, intercalare consuerunt. Quod si quis calculatorum facere negligens CCC

---

3. Bissextus ys forþon geciged, forþon bis ys twia 7 sextus ys se syxta, forþam we cweðað on þam geare nu to-dæg: 'VI. kl. Martii' 7 eft on morgen '.VI. kl.' *Byrhtferð, p. 64.*
4–8. *Cp. DT. x.*
5. *Cp. iv. 17 above and viii. 6 below.*—Facit enim ratio quadrantis (quem bissextum uocant), ut sol aequinoctialem sui cursus locum in signifero circulo, nunc in primo exortu suo, nunc in meridie, nunc in occasu, nunc media nocte comprehendat. *Bede, Epistola ad Wicredam, de paschae celebratione, siue de aequinoctio uernali iuxta Anatolium (Giles i. 156).*—Þæt ys eac to witanne þam þe þis wyle cunnan mid gesceade, þæt synd twelf tacna, 7 on þissum twelf tacnum wunað seo sunne þreohund daga 7 fif 7 syxtig daga 7 syx tida. On þam forman geare þe bissextus byð, gæð seo sunne on ærne mergen on þæt tacen þe ys Aries, þæt ys ram genemned, 7 þæs oðres geares heo gæð on þæt tacen on middum dæge, 7 þy þriddan heo sihð to þam tacne oð æfen, 7 on þam feorðan geare heo ferecað on middre nihte to þam foresprecenan tacne, 7 swa on þam fiftan geare he becymð to þam stede þe heo ærest geset wæs on frimðe þisre worulde. *Byrhtferð, pp. 68–70.*— *Cp. Byrhtferð, pp. 114–16.*

on middæge · þriddan geare on æfen · feorðan geare on middere nihte · on ðam fiftan geare eft on ærne merigen;

6. Witodlice ælc ðæra feower geara agifð six tida · *þæt* sind feower 7 twentig tida · An dæg · 7 an niht;

7. Þone dæg setton Romanisce witan to ðam monðe þe we hatað februarius · forðan ðe se monað is ealra scyrtst · 7 endenext;

8. Be ðam dæge spræc se wisa Augustin*us* · *þæt* se ælmihtiga scyppend hine gesceope fra*m* frimðe middaneardes · to micelre gerynu · 7 gif he bið forlæten unateald · þærrihte awent eal þæs geares ymbrene ðwyres · 7 hê belimpð ægðer ge to ðære sunnan ge to ðam monan · forðan þe ðær is ân dæg · 7 ân niht;

9. Gif ðu nelt hine tellan eac to ðam monan swa swa to þære sunnan · ðonn*e* awægst ðu þone easte*r*lican regol · 7 ælces niwan monan gerîm ealles þæs geares;

---

5 (*cont.*) A. middæge · þriddan geare on *desunt*   E. middege
C. þriddan on *for* þriddan geare on   C. feorðan on *for* feorðan geare on
ABDE. middre, C. midre, F. mi`d´dre   C. on ðam *desunt*   D. þan
A. ærne merne mergen   C. morgen, D. merien, F. mergen
6. A. þære, E. þare, F. þara   A. feowra   A. geare
BDEF. agyfð   BDF. syx   A. sindon, BDE. synd   E. tida
þæt is an *for* tida · An   AF. on dæg 7 on niht   BCDE. an *for*
An   E. deg
7. A. þæne   E. deg   B. settan   C. þa Romaniscan
A. witon   B. weras 7 witan   DE. on *for* to   C. febryarius
ADF. forþam, B. forðam   C. ðe *for* ðe se   A. monoþ   A. ys
C. ealre, F. eallra   A. scyrtost, C. sceortast, DE. scyrtest, F.
scirtst   A. endeneht, B. endenyhst, C. endenexst, F. ændenihxt
8. E. dege   A. sprycþ, C. sprec, E. sprecð   C. ðe   E.
wise   C. agustinus   C. ðe   F. æ`l´myhtiga   A. hyne
A. gesceop   A. frymþe, BCDE. frymðe   D. middangeardes
A. geryne   E. 7 *deest*   BEF. gyf   A. byþ, B. byð   AC.
ungeteald, BDEF. unteald   F. þærryhte   ABCDF. eall
AC. ymbrynu, B. ymbryn, F. ymbbryn, DE. ymbren   A. he eft
gelimþ ge ægþer ge   E. egðer   C. þan   DF. forþam, E.
forðam   A. ys   E. deg   A. ane niht
9. BF. Gyf, E. gyf   A. hyne   B. awæst   A. þonne *for*
þone   E. eastorlican   B. niwes   E. mona

solum ac LXV diebus omnes se annos agere debere putauerit, magnum sibi mox inueniet annui circuitus occurrisse dispendium. *DTR. xxxviii.*

7. Ob hoc autem bissexti diem in mense Februario placuit intercalare Romanis, quia hic breuior caeteris et extremus anni mensis erat. *DTR. xl.*

8. Uerum si quis nostram in hoc spernendam putat assertionem, legat beati Aurelii Augustini quartum de sancta trinitate libellum, ubi de senarii numeri . . . perfectione disserens, etiam huius quadrantis mentionem facere, imo et eum certi mysterii gratia per omnipotentem conditoris sapientiam prouisum factumque docere non omisit. *DTR. xxxix.*

8-9. Inter haec autem meminisse debet calculator, ut lunam mensis Februarii caeteris annis XXIX dierum, anno autem bissextili computet XXX siue illam ante intercalatum quadrantem, seu postmodum terminari contingat . . . Palam namque apparet quadrantem, de quo sermo est, non ad solis tantummodo, sed ad lunae cursum aeque pertinere: quia si

---

7. Gyt we eow cyðað ymbe þone arwyrðan bissextum. Forþon Romani hine gelogodon on þissum monðe, þæt ys on Februario, forþam he ys scyrtest ealra monða 7 se ytemesta æfter Lydenwarum. *Byrhtferð, p. 64.*

8. He (sc. bissextus) wæs on frymðe þisre worulde. *Byrhtferð, p. 68.*—Ac ærest ys to witanne þæt se mona sceal habban his bissextum, þæt ys þæt he hæfð þy geare anum dæge ma daga þonne þy oðrum geare. Februarius hæfð lunam uigessimam nonam þreo gear, 7 þonne he sceal habban þrittig nihta ealdne monan. *Byrhtferð, pp. 62-4.*

8-9. Betwux þisre spræce sceal se rimre geþencean þæt he gedó þæt Februarius monð þy geare hæbbe þrittig nihta ealdne monan, beo he þær þær he beo nigon 7 twentig nihta eald geendod, gif he byð beforan þam intercalatum uel interpolatum diem, þæt ys gif he byð beforan þam gesettan dæge, oððe he byð betwynan þam gelogodan dæge, þæt ys bissextus. . . . Þæt ys wislice to witanne, þæt bissextus, þe we ymbe synd sprecende, þæt he gebyrað ægðer ge to þære sunnan ryne ge tó þæs monan. Soðlice gif man forwyrnð þam worigendan monan (his) quadrans, þæt ys lunam .XXX. þanne byð mycel gedwyld on þam Easterlican termene; þæt gewyrð þæt man hæfð lunam .XIIII. anum dæge ær his riht þegenscipe sý. *Byrhtferð, pp. 70-2.*

[VIII.] DE SALTU LUNE.

1. Swâ swâ þære sunnan sleacnys acenð ænne dæg · 7 ane niht æfre ymbe feower gear · swa eac þæs monan swyftnys awyrpð ût ænne dæg · 7 ane niht of ðam getele his rynes · æfre embe nigontyne gear · 7 se dæg is gehâten Saltus lune · þæt is ðæs monan hlyp · forðan ðe he oferhlypð ænne dæg · 7 swa near þam nigonteoðan geare swa bið se 'niwa' mona braddra gesewen;

2. Se mona wæs æt fruman on æfen gesceapen · 7 æfre siððan on æfen · his ylde awent;

VIII. ACE. *title deest*   B. DE SALTV LVNEA   D. lunę

1. A. sunnan *deest*   CF. sleacnes, E. sleacnyss   A. anne
E. deg   A. ane *deest*   E. ymb, F. embe   A. swyftnyss,
BF. swiftnes, C. swyftnes   A. awyrþ   E. deg   BCF.
getæle   AB. hys   F. rines   ABCDE. ymbe   AE.
nygontyne, B. neogontyne, F. nigentine   C. þe *for* se   E. deg
A. ys   E. lunę   A. ys   ADF. forþam   A. anne   E.
deg   F. neor   B. neogonteoðan, F. neogontǽoþan   A. byþ,
BC. byð   C. ðe   C. niwe   C. braddre
2. *In* E, *the leaf (fol. 142) is trimmed, and from this point (line 11) on the text is affected. I note all last words of lines that are mutilated. Lost letters are supplied in square brackets.*
E. Seo *for* Se   ABC. syððan, D. syþþan, E. syðða[n], F. syþðan
A. hys   C. ylde he awent   F. awænt

lunae quoque quadrantem accommodare negaueris, sed bissextili anno eiusdem quantitatis mensem lunarem Februario, cuius et antea solebas, aptaueris, fit profecto ut et quartadecima luna paschalis eiusdem anni, pridie quam debuerat adueniat. Ideoque paschalis ratio uacillet, et totius mox anni cursus titubet, statusque ille semper inuiolabilis circuli decennouenalis magis magisque turbatus euertatur. *DTR. xli.*

1. Sicut, inquam, saltum lunae quem dicunt, locus et hora citior incensionis eius per X et IX annos efficit: ita e contrario bissextum non alia causa, quam tarditas solaris cursus generat. *DTR. xxxviii.*—Notandum sane, quod huius ratio saltus lunaris, longa sui facit exundantia crementi lunam aliquoties maiorem quam putatur[1] uideri, adeo ut etiam die tricesima uesperascente illam non gracilem in coelo apparere contingat, et quanto circuli decennouenalis terminus amplius instat, tanto hoc crebrius patiatur causa existente perspicua, quod saltus ille, de quo loquimur, iam maxima tunc sit ex parte perfectus. *DTR. xliii.*

2. Quia nimirum luna quae uespere primum mundo exorta est uespertinis ex eo semper horis necesse est aliam aliamque sumat aetatem. *DTR. xliii.*

[1] *Should probably be* computatur.

1–2. We cwædon ǽr, eall swa þære sunnan slecnys acenð anne dæg 7 ane niht æfre binnan feower wintrum, swa eac þæs monan swyftnys awyrpð út anne dæg 7 ane niht æfre ymbe nigontyne gear, 7 se dæg ys gehaten saltus lunę, þæt ys þæs monan hlyp, forþan he oferhlypð anne dæg . . . Swa near þam nigonteoðan geare, swa byð se niwa mona bradra gesewen. Se mona wæs æt fruman on æfen gesceapen, 7 æfre syððan on æfen byð his ylde awend. *Byrhtferð, pp. 76–8.*

2. *Cp. ii. 3 above.*

3. Gif hê bi∂ ær æfene fra*m* ∂ære sunnan geedniwod · hê
biδ þonn*e* sona · æft*er* sunnan setlunge niwe geteald;

4. Gif hê ∂onn*e* æft*er* sunnan setlunge ontend bi∂ · oþþe
on middere nihte · o∂∂e on hancrede · ne bi∂ he næfre niwe
geteald · þeah ∂e hê hæbbe þreo 7 twentig tida · ær ∂an ∂e hê
becume to ∂an æfene þe hê on gesceapen wæs;

5. Be ∂isu*m* is oft micel embspræc · þonn*e* ∂a læwedan
wylla∂ habban ∂one monan · be ∂an þe hî hine geseo∂ · 7
∂a gelæredan hine healda∂ · be ∂isu*m* foresædan gesceade;

f. 260 r. 6. Hwilon | bi∂ se môna ontend of þære sunnan on dæg ·
hwilon on niht · hwilon on ærne merigen · hwilon on æfen · 7
swa mislice · ac he ne bi∂ þeah niwe · ær ∂an ∂e hê þone æfen
gesih∂;

7. Ne sceal nân cristenman · nan ∂ing be ∂am monan
wîglian · gif he hit de∂ · his geleafa ne bi∂ naht;

---

3. BEF. Gyf   BF. by∂   BC. æfenne   AB. by∂   A.
þænne   F. æfter þære sunnan   A. setle unniwe *for* setlunge
niwe   BDE. setlgange
4. BEF. Gyf   A. þænne   ABF. by∂   E. o∂∂[e]   DE.
middre   A. on *deest*   F. hancræde   A. byþ, BCF. by∂
E. gete[ald]   EF. þeh   C. he *deest*   BC. habbe   ACDEF.
ær þam   C. ∂æ   ABCDEF. þam   BCF. æfenne, E. æfen[e]
B. gescapen
5. BD. þysum, F. þissum   A. ys   ABCF. oft *deest*   BF.
mycel   AD. ymbespræc, BCEF. ymbspræc   E. þo[nne]   A.
∂a *deest*   BEF. willa∂   C. þon *for* ∂one   ABD. þam, F.
þann   F. hy   CD. læredan, E. gelaredan, F. gelæredon   A.
hyne   D. þysum   F. foresædum   E. gesc[eade]
6. A. byþ, B. by∂   C. ∂e   EF. ontent   B. hwilon on
æfen · hwilon on ærne merigen   AC. ærne *deest*   AF. mergen,
C. morgen   B. mistlice, E. mislic[e]   A. byþ na þeah
B. by∂   C. nipe *for* niwe   ADE. þam   A. þæne   A.
gesyhþ, C. gesyh∂
7. ABDF. -mann   AC. þing wiglian be þa*m* monan   B.
þincg   BEF. gyf   A. hyt, F. hit *deest*   A. hys   A.
by∂   D. nis *for* ne bi∂

Sources and Parallels 61

3–4. Et siquidem eam (sc. lunam) paulo ante uesperam accendi a sole contigerit, mox sole occidente primam computari et esse necesse est . . . Sin autem post occubitum solis accendatur, non tamen primam priusquam uesperam uiderit, sed tricesimam potius oportet aestimare. Etiam si XXIII horas post occasum solis accensa suppleuerit, illam tamen quam occidente sole habuerat, ne primae conditionis ordo turbetur, usque ad alium eius occasum retinere debebit aetatem. *DTR. xliii.*

5. Liquet itaque quia uetus haec de hac aetate lunari quaestio est, etiam olim beati Papae Leonis industria diligenter exquisita, ipsa est quae longam inter orientis et occidentis ecclesias grauemque controuersiam fecit. *DTR. xliii.*

---

6. *Cp. vii. 5 above.*
7. Nu wîgliað stunte men menigfealde wîgelunga on ðisum dæge (*i.e. January 1st*), mid micclum gedwylde, æfter hæðenum gewunan, ongean heora cristendom, swylce hî magon heora lîf gelengan, oþþe heora gesundfulnysse, mid þam ðe hî gremiað þone Ælmihtigan Scyppend. Sind eac manega mid swa micclum gedwylde befangene, þæt hî cepað be ðam monan heora fær, and heora dæda be dagum, and nellað heora ðing wanian on monan-dæg, for anginne ðære wucan. *Homilies, i. 100.*—Nis þæs mannes cristendom naht, þe mid deoflicum wîglungum his lîf adrihð. *Ibid., p. 102.*

8. Swa lengra dæg swa bið se niwa mona ufor gesewen
7 swa scyrtra dæg swa bið se niwa mona nyðor gesewen;

9. Gif seo sunne hine onælð ufan · þonne stupað hê · gif heo
hine onælð riht ðwyrs · þonne bið hê emlice gehyrned · gif
heo hine ontent neoðan · þonne capað he upp · forðan ðe hê
went æfre ðone hricg · to ðære sunnan weard;

10. He bið swa awend · swa swa heo hine atent;

11. Nu cweðað sume men þe ðis gescead ne cunnon þæt se
mona hine wende be ðan ðe hit wedrian sceall on ðam monðe ·

---

8. E. deg   AB. byð   A. niwa *deest*, C. niwe   A. ufror,
C. ufer   F. gesawon   E. deg   AE. byþ, B. byð   B.
nyðer, C. neoðer, D. neoðor, F. niðor
9. BE. Gyf, F. gyf   A. hyne   CDE. ufon   A. he · 7 gif,
F. he 7 gyf   BE. gyf   A. hynne   F. ryhte   ABCDF.
þwyres, E. þwyre   F. þænne   ABF. byð   B. ymlice   E.
geherned   A. gehyrned 7 gif   BEF. gyf   E. he *for* heo
A. hyne   BC. ontend   A. nyþan   BCDE. up   AD.
forþam, C. forþon, E. forðam   A. þæne   CE. ricg, F. hrigc
A. on *for* to   E. sunna
10. AB. byð   B. onwend, D. gewend, E. awent   B. seo
sunne *for* heo   A. hyne, E. hi[ne]   ACDEF. ontent, B. ontend
11. ABDEF. menn   C. cunnen   E. mo[na]   A. hyne
F. wænde   ACDEF. þam   A. he *for* hit   B. wuderian,
C. wederian, F. widrian   ABCDEF. sceal

8. Quo dies longior, eo sit noua luna excelsior: et quo breuior atque ad meridiem decliuior est dies, eo deiectior noua luna cernatur. *DTR. xxv.*

9. Cum ergo die crescente sol a meridianis plagis ad boreales paulatim partes ascenderit, necesse est luna, quae eo tempore nata est, ociori transitu solem ad borealia signa praecurrat: atque ideo cum noua post occasum solis uidetur, quae ad septentrionem solaris occasus occasura est, nimirum non iuxta, sed supra solem sita est, quo inferiora eius illustrante, aequalia pene cornua protendere, et instar nauis supina ire uidetur. At reuerso post solstitium aestiuum ad inferiora et australia cursu solis, luna quoque illis nata mensibus, ad inferiora cursum tendat necesse est: unde fit, ut quae ad australem partem solis, qui occiderat occasura est: absque ulla dubietate cum primo post occasum solis apparet, non iam supra illum, sed iuxta illum ad meridiem posita uideatur. Atque ideo aquilonalia eius latera sole aspectante cernuntur erecta progredi: semper enim luna auersis a sole cornibus, rotundam sui partem pandit ad illum. *DTR. xxv.*

10. Lunae autem status idem, eademque sit pro uariante solis digressu conuersio. *DTR. xxv.*

11. Et inde uulgi creuit opinio, lunam cum supina et celsior incedit, turbines tempestatum: cum uero erecta, et in austros deiectior, tranquillitatem designare . . . Non ergo lunae conuersio, quae naturalis est et fixa, potest futuri mensis portendere statum. *DTR. xxv.*

8–10. *Cp. DNR. xx.*
9. *Cp. iii. 12 above.*

ac hi[ne]¹ ne went næfre naðor · ne weder · ne unweder · of ðam ðe his gecynde is;

12. Men magon swa ðeah · þa ðe fyrwite beoð cepan be his bleo · 7 be ðære sunnan · oððe þæs roderes · hwilc weder toweard bið;

13. Hit is gecyndelic · *þæt* ealle eorðlice lichaman · beoð fulran on weaxendu*m* monan · þonn*e* on wanigendum;

14. Eac ða treowu þe beoð aheawene on fullu*m* monan · beoð heardran · wið wyrmǽtan · 7 langfærran · þonn*e* ða þe beoð on niwu*m* monan aheawene;

¹ G. hit

---

11 (*cont.*) A. hyne, G. hit *for* hine   F. awent   A. naðor
deest   A. hys, B. him *for* his   C. gecynd   AB. ys
12. ABDEF. Menn   F. þeh   B. þa þa   A. fyrwyte, B.
fyrwytte   AC. kepan   A. hys   AB. hwylc   A. byþ,
BC. byð
  The second fragment of E *ends here*
13. A. hyt   AF. ys   F. gecindelic   F. fullran   A.
weaxenda, C. weaxend*e*   A. þænne   A. waniendan, CDF.
waniendum
14. A. Ac   AC. treow, BF. treowa   C. byð   F. wirm-
ætam   B. lengfærran, CF. langferran   A. þænne, B. þonnie

12. Sed qui curiosi sunt huiusmodi rerum, coloris uel eius (*sc.* lunae), uel solis, uel coeli ipsius . . . saepe statum aeris, qui sit futurus explorant. *DTR. xxv.*
13. Nam et defectui eius (*sc.* lunae) compatiuntur elementa, et processu eius quae fuerint exinanita cumulantur ut animantium cerebra maritimorum humida: siquidem pleniores ostreae reperiri ferantur multaque alia, cum globus lunaris adolescit. *DTR. xxviii.*
14. De arborum quoque internis idem allegant qui hoc usu proprio compererunt. Haec beati Ambrosii uerba etiam architectorum omnium ars, et quotidianus usus adfirmat, qui obseruandum praecipue docent, ut a quintadecima luna usque ad uicesimam et secundam arbores praecidantur, ex quibus uel liburnae texendae, uel publica quaeque sunt opera facienda. His enim tantum octo diebus caesa materies immunis seruatur a carie, reliquis autem diebus praecisa etiam in

---

12. *Cp. Isidore, DNR. xxxviii, Bede, DNR. xxxvi and Comm. in Lib. Gen. (Giles vii. 15).*
13. Cuius (*sc.* lunae) etiam augmentis decrementisque, mira quadam Prouidentiae arte, omne quod gignitur, alitur atque crescit. *Isidore, DNR. xix. 2.*
13–15. Is hwæðere æfter gecynde on gesceapennysse ælc lichamlice gesceaft ðe eorðe acenð fulre and mægenfæstre on fullum monan þonne on gewanedum. Swa eac treowa, gif hî beoð on fullum monan geheawene, hî beoð heardran and langfærran to getimbrunge, and swiðost, gif hî beoð unsæpige geworhte. Nis ðis nan wîglung, ac is gecyndelic ðincg þurh gesceapenysse. Hwæt eac seo sæ wunderlice geþwærlæcð þæs monan ymbrene; symle hî beoð geferan on wæstme and on wanunge. And swa swa se mona dæghwonlice feower pricon lator arist, swa eac seo sæ symle feower pricum lator fleowð. *Homilies i. 102.*—Ða gleawe sægengan wel hig understandað þæt eorðlice lichaman beoð fulran on weaxendum monan þonne on wanigendum. Eac þa treowa þe beoð aheawene on fullum monan beoð heardran wið wyrmætan 7 langferran þonne þaþe beoð on niwum monan aheawene. Seo sæ 7 se mona geþwærlæcað heom betweonan. Æfre hig beoð geféran on wæstme 7 on wanunge; 7 swa swa se mona dæghwamlice feower prican lator arist þonne he dyde on þam oðrum dæge, swa eac seo sæ symle feower prican oððe fif lator flowð. *Byrhtferð, pp. 156–8.*

15. Seo sæ 7 se mona geðwærlæcað him betweonan · æfre
hî beoð geferan on wæstme · 7 on wanunge · 7 swa swa se
môna dæghwomlice feower pricon lator arîst · ðonne hê on
ðam oðrum dæge dyde · swa eac seo sæ symle feower pricon
lator flewð;

IX. DE DIUERSIS STELLIS.

1. Sume men cweðað *þæt* steorran feallað of heofenum · Ac
hit ne sind nâ steorran *þæt* ðær feallað · ac is fyr of ðam rodore · þe sprincð of ðam tunglum swa swa spearcan doð of fyre;
2. Witodlice swa fela steorran sind gyt on heofenum swa
swa on frymðe wæron · þa ða hî God gesceop;
3. Ealle mæst hî sind fæste on ðam firmamentum · 7 ðanon
ne afeallað ða hwile þe ðeos woruld stent;
4. Seo sunne · 7 se môna · 7 æfensteorra · 7 dægsteorra · 7
oðre ðry steorran ne sind na fæste on ðam firmamentum · ac
habbað heora agenne gang on sundron;

---

15. C. þeo   C. þe   A. geþwærlæceaþ, F. geþwærlæceað, C.
gehwærlæcað   A. betwynan   A. hig   C. byð   B. wæsme
C. waniunge   A. swa *for* swa swa   ABCDF. dæghwamlice
ABF. prican   F. later   A. þænne   C. odrum, F. oðran
F. dæge ær dyde   A. simble, C. æfre *for* symle   B. prican
CF. fleowð

IX. AC. *title deest*
1. ABDF. menn   C. on *for* of   AF. heofonum, B. heofenan
AF. synt, BD. synd   C. fællað   A. ys   A. on *for* of
DF. rodere   BCDF. springð   B. tunglon
2. AF. feala   A. synt, BDF. synd   F. heofonum   A.
hig, D. hy   A. gescop
3. F. mæste   A. sindon, BCF. synd   C. ðanon *deest*   CF.
na   F. feallað   A. hwyle   F. stænt
4. F. þreo   A. syndon, BDF. synd   B. hyra, C. heore,
F. hyre   CF. sundran

eodem anno, interna uermium labe exesa in puluerem uertitur.
DTR. xxviii.

15. Maxime autem prae omnibus admiranda tanta oceani
cum lunae cursu societas . . . Sicut enim luna, iuxta quod et
supra docuimus, IIII punctorum spacio quotidie tardius
oriri, tardius occidere quam pridie orta est uel occiderat solet,
ita etiam maris aestus uterque, siue diurnus sit, et nocturnus,
seu matutinus, et uespertinus eiusdem pene temporis interuallo tardius quotidie uenire, tardius redire non desinit.
DTR. xxix.

1-3. Falsa autem opinio et uulgaris est, nocte stellas
cadere, cum sciamus ex aethere lapsos igniculos ire per
coelum, portarique uentis, uagique lumen sideris imitari;
stellas autem immobiles fixasque permanere in coelo. *Isidore, DNR. xxv. 1.*

4. Inter coelum terramque septem sidera pendent, certis
discreta spatiis, quae uocantur errantia, contrarium mundo

---

15. *Cp. DNR. xxxix.*
1. Þæt manega menn geseoð feallan of þære heofone swylce hyt sýn
steorran, hyt beoð spearcan of þam rodere þurh þæs windes blǽs,
þe þænne swyðlice þa heannyssa þæs roderes secð mid his þodenum.
Byrhtferð, p. 128.
1-4. Stellae lumen a sole mutuantes, cum mundo uerti, utpote in
uno loco fixae, et non stante mundo, uagae ferri dicuntur: exceptis
iis quae planetae, id est, errantes uocantur: easque diei aduentu
celari: nec unquam coelo decidere, fulgor plenilunii, et solis probat
deliquium. Quamuis uideamus igniculos ex aethere lapsos portari
uentis, uagique lumen sideris imitari, trucibus cito coorientibus
uentis. *DNR. xi.*—Ne standað na ealle steorran on ðam steapan
rodere ac hi sume habbað synderlicne gang beneoðan ðam ródere
mislice geendebyrde, 7 ða ðe on ðam rodere standað tyrnað æfre
abutan mid ðam bradan rodere on ymbhwyrfte ðære eorðan, and
heora nan ne fylð of ðam fæstan rodere ða hwile ðe ðeos woruld
wunað swa gehal. *Hexameron, 229-35.*
2. *Cp. i. 6 and v. 2 above.*
4. Heora (*sc.* planetarum) ælc gæð on his agenum ryne hwilon ufor,
hwilon nyðor and ne synd na fæste on ðære rodorlican heofonan swa
swa oðre tunglan. *Interrogationes, xxi.*

5. Þa seofon sind gehâtene Septem planete · 7 ic wât þæt
hit wile þincan swiðe ungeleaffullic ungelæredum mannum ·
gif we secgað gewislice be ðam steorrum · 7 be heora gange;
6. Arcton hatte ân tungel on norðdæle · se hæfð seofon
steorran · 7 is forði oðrum naman gehâten Septemtrio · þone
hâtað læwede men Carles wæn · se ne gæð næfre adune under
ðissere eorðan · swa swa oðre tunglan doð · Ac he went
abutan hwilon up · hwilon adûne · ofer dæg · 7 ofer niht; |

f. 260 v.   7. Oðer tungel is on suðdæle þisum gelic · ðone we ne
magon næfre geseon;

8. Twegen steorran standað eac stille · an on suðdæle · oþer
on norðdæle · ða sind on leden Axis gehatene · þone suðran
steorran we ne geseoð næfre · þone norðran we geseoð · þone
hatað men Scipsteorra;

9. Hî sind gehatene axis · þæt is · ex · forðan ðe se fir-

---

5. B. seofan, F. seofone   A. syndon, BCDF. synd   A.
gehatenne   B. planetȩ, F. plante   A. hyt   AD. wyle
AF. þincean   B. swyðe, C. swiðe *deest*   CF. ungeleafullic
BF. gyf   A. secgeaþ   B. steorran   BF. hyra
6. B. Arhcton, C. Areton, F. Arhton   F. sum *for* an   B.
tungol   C. þe *for* se   A. ys   C. gehatan   A. septentrio
AF. þæne   ABDF. menn   F. se se ne gæð   CF. adun   A.
þisse, B. þyssere   A. tungla   CDF. abuton   B. adune · 7
hwilon up *for* abutan hwilon up · hwilon adune   A. upp   A.
hwilon *deest*   CF. adun
7. AF. ys   BD. þysum, F. þissum   A. þænne
8. DF. an *deest*   A. synt, BDF. synd   A. þæne   A.
þæne   A. þæne   ABDF. menn   ACF. scyp-
9. A. synt, BDF. synd   A. ys, D. i *for* is   ABDF. forþam

agentia cursum, id est, laeuum, illo semper in dextram praecipiti. *DNR. xii.*

5. *DNR. xii and xiii passim.*

6. Arcturus est ille quem Latini Septentrionem dicunt, qui septem stellarum radiis fulgens, in seipso reuolutus rotatur, qui ideo Plaustrum uocatur, quia in modum uehiculi uoluitur, et modo tres ad summa eleuat, modo quatuor inclinat. Hic autem in coeli axe constitutus semper uersatur, et nunquam mergitur. Sed dum in seipso uoluitur, et nox finitur. *Isidore, DNR. xxvi. 3.*

6–7 Denique ipsos Septentriones, qui nobis supra uerticem ascendunt, neque occidunt unquam, non cernit Troglodytice, et confinis Aegyptus. Porro ipsorum sidus permaximum, et Dei quondam nomine cultum, non solum nos Britanni, sed nec Italia quidem potest uidere Canopum. *DTR. xxxii.*

8. Bootes, stella est quae Plaustrum, id est Septentrionem, sequitur, qui etiam ab antiquis Arctophylax dicitur, siue minor Arctos. Unde et quidam eam Septentrionem dixerunt. Hanc spectant praecipue qui nauigare noscuntur. *Isidore, DNR. xxvi. 5.*

8–9. . . . cuius (*sc.* coeli) uertices extremos, circa quos sphaera coeli uoluitur, polos nuncupant, glaciali rigore tabentes. Horum unus ad septentrionalem plagam consurgens Boreas: alter deuexus in Austros, terraeque oppositus, australis uocatur. *DNR. v.* Duo sunt autem, ut diximus, poli, quibus coelum uoluitur, Boreus, quem Aquilonium uocamus. Hic Arcti, id est, septentriones, qui nobis semper apparent. Cui contrarius est Notius, qui australis dicitur. Hic est, qui terra, ut ait Cicero, tegitur, et ἀφανὴς a Graecis nominatur. *Isidore, DNR. xii. 6.*

6–7. *Cp. DTR. xxxiv (Giles vi. 215–16) and DNR. xlvi.*
9. *Cp. i. 5 above.*

mamentum went on ðam twam steorrum · swa swa hweowul
tyrnð on exe · 7 forði hî standan symle stille;
10. Pliâde sind gehatene ða seofon steorran · þe on hærfeste upagað · 7 ofer ealne winter scinað · gangende eastan
westweard;
11. Ofer ealne sumor hî gað on nihtlicere tide under þyssere
eorðan · 7 on dæg bufon;
12. On winterlicere tide hî beoð on niht uppe · 7 on dæg
adune;

13. Comête sind gehatene þa steorran ðe færlice 7 unge-
wunelice æteowiað · 7 sind geleômode swa þæt him gæð of se
leôma swilce oðer sunbeam · hî ne beoð na lange hwile
gesewene · ac swa oft swa hî æteowiað · hî gebicniað sum ðing
niwes toweard · þære leode ðe hî ofer scinað;
14. Þeah ðe we swiðor sprecon be heofenlicum tunglum ·
ne mæg swa ðeah se ungelæreda leornian heora leohtbæran
ryne:—

---

9 (cont.) BF. steorran   B. hweogel, F. hweogul   D. tyrþ
B. eaxe   A. standaþ, BCDF. standað   A. symble, C. æfre
for symle
10. C. Pliades   A. synt, BDF. synd   C. herfeste   C.
upgað, F. upgæð   A. ofor   F. eallne   F. scynað
11. C. sumur   A. nihtlic'e're, BD. nihtlicre, F. nihtli'c're   A.
þisse, BCF. þissere   AB. bufan
12. BCDF. winterlicre
13. A. syndon, BCDF. synd   CF. æteowað   ABDF. synd
A. gelomode, B. geleomade   A. swa þæt se leoma ys swilce oþer
sunnbeam for swa—sunbeam   B. swylce   B. sunnbeam
A. swa hig   C. æteowað, F. ætypiað   A. hig   C. gebycniaþ
F. þincg   F. þare   F. scynað
14. A. swyþor   B. sprecað, C. sprecan   A. ungelærede
BF. hyra   C. leoht be þam for leohtbæran

10–12. Pleiades sunt multae iuges stellae quas etiam
botrum appellamus a multitudine stellarum. Nam et ipsae
septem esse dicuntur, sed amplius quam sex nullus conspicere
potest. Haec ab oriente surgunt, et appropinquante diei
claritate stellarum eius ordo distenditur . . . Has Latini Uer-
gilias appellauerunt, eo quod uere oriantur, et eo magis
caeteris praedicantur, quod his exorientibus aestas signi-
ficatur, occidentibus hiems ostenditur, quod aliis penitus non
est traditum signis. *Isidore, DNR. xxvi. 6.*

13. Cometae sunt stellae flammis crinitae, repente nas-
centes, regni mutationem, aut pestilentiam, aut bella, uel
uentos, aestusue portendentes . . . Breuissimum quo cerne-
rentur spatium septem dierum annotatum est, longissimum
LXXX. *DNR. xxiv.*

10–12. Se sumor hafað hundnigontig daga; þonne gangað þa seofon steorran on uhtan ûpp 7 on æfen on setl. Se winter hafað tu 7
hundnigontig daga, 7 þonne gongað þa seofon steorran ûp on æfen 7
on dægered on setl. *Martyrologium, pp. 80 and 202.*

13. Cometes uero stella quae temporibus congruis domini nutu
apparet, ideo uocata est quod comas luminis ex se fundat, et cum
apparuerit aut pestilentiam, aut famem, aut bella significat, excidium
etiam patriae mortemque principum. *Excerptio Abbonis ex Higino de
Figuratione Signorum. MS. Trinity College Cambridge R. 15. 32,
p. 212.*—Sind eac sume steorran leohtbeamede, færlice arîsende, and
hrædlice gewîtende, and hî symle sum ðing nîwes mid heora upspringe
gebîcniað. *Homilies, i. 610.*—An steorra ys genemned cometa.
Þonne he ætywð, þonne getacnað he hungor oððe cwealm oððe
gefeoht oððe tostencednyss þæs eardes oððe egeslice windas. *Byrhtferð, p. 132.*

[X.] DE DUODECIM UENTIS.

1. Þeos lyft þe we on lybbað is an ðæra feower gesceafta ·
þe ælc lichamlic ðing on wunað;

2. Feower gesceafta sind · þe ealle eorðlice lichaman on
wuniað · þæt sind · Âer · Ignis · Terra · Aqua;

3. Aer · is lyft · Ignis · fyr · Terra · eorðe · Aqua · wæter;

4. Lyft is lichamlic gesceaft swiðe þynne · seo ofergæð
ealne middaneard · 7 upastihð fornean oð þone monan · on
ðam fleoð fugelas · swa swa fixas swymmað on wætere;

5. Ne mihte heora nan fleon · nære seo lyft ðe hî berð;

6. Ne nan man ne nyten næfð nane orðunge · buton ðurh
ða lyfte;

7. Nis na seo orðung ðe we utblawað · 7 innateoð ure
sawul · ac is seo lyft þe we on lybbað on ðisum deadlicum
life;

8. Swa swa fixas cwelað gif hî of wætere beoð · swa eac
cwelð ælc eorðlic lichama · gif he bið þære lyfte bedæled;

X. AC. *title deest*
1. A. Þis lift   A. libbaþ, BCF. libbað   ABF. ys   C.
an—on wunað *desunt*   F. þara   A. feorþa, BF. feorða, *for* feower,
D. feower *deest*   A. þæ   A. þingc
2. A. feowor   A. syndon, BDF. synd   A. þæt ys, F. þæt is
for þæt sind   BCD. synd   A. lyft · 7 fyr · 7 eorþe · 7
wæter *for* Aer · Ignis · Terra · Aqua
3. A. *the whole paragraph deest*   C. Ignis is fyr
4. AF. ys   A. swyþe, BF. swyðe   C. þeonne, F. þinne
D. middangeard   A. uppastihð, CF. upastyhð   C. forneah   AF.
þæne, C. ðene   A. fleogað   F. fyxas   BF. swimmað, C.
swimmaþ
5. F. myhte   F. hyra   A. þæt *for* seo   A. byrþ,
BDF. byrð
6. ABDF. mann   C. butan   B. þæt *for* ða   BCDF. lyft
7. A. Nys   C. þeo *for* seo   BCDF. inateoð   AC. saul   A.
ys   C. þy *for* seo   A. libbaþ, BCF. libbað   A. þisum, B.
ðyssum, D. þysum   B. deadlican   F. lyfe
8. AF. fyxas   F. cwellað   BF. gyf   A. heo *for* hi   B.
beað   AF. cwylð   ABF. gyf   ABF. byð   A. lifte
A. bedæld

1–3, 6. Mundus est uniuersitas omnis, quae constat ex coelo et terra, quatuor elementis in speciem orbis absoluti globata: igne, quo sidera lucent: aere, quo cuncta uiuentia spirant: aquis, quae terram cingendo et penetrando communiunt: atque ipsa terra. *DNR. iii.*

4, 7, 13. Aer est omne quod inani simile uitalem hunc spiritum fundit, infra lunam, uolatus auium nubiumque, et tempestatum capax. *DNR. xxv.*

---

1. On ðam lyfte we lybbað ealle. *Hexameron, 412.*
2. 'On hû fela gesceaftum stent þes middaneard?' 'On fêowrum: on fŷre and on lyfte, on wætere and on eorþan.' *Interrogationes, xviii.*—On ælcum lichamlicum gesceafte syndon feower ðing, eorðe and wæter, fyr and lyft. *Hexameron 405 sq.*—Eac ys þam preoste to witanne þæt þes middaneard stent on feower gesceaftum underwryðed, and eac mid feower mægenþrymmum gefrætwod. Þa feower gesceaft synd þus geciged: aer, ignis, aqua, terra. *Byrhtferð, p. 92.*— Totus homo interior et exterior septem qualitatibus constat: Interior: intellectu, memoria et uoluntate; exterior: terra, aqua, igne, aere. *Ibid., p. 230.*
4–6. Ðæt lyft he gesceop to ures lifes strangunge, ðurh ðæt we orðiað and eac ða nytenu, and ure fnæst ateorað gif we ateon ne magon mid urum orðe into ús ðæt lyft and eft utablawan ða hwile ðe we beoð cuce. Ðæt lyft is swa heah swa swa ða heofonlican wolcnu and éac eal swa brad swa swa ðære eorðan bradnyss. On ðære fleoð fugelas ac heora fiðera ne mihton na hwider hi aberan gif hi ne abære seo lyft. *Hexameron, 130–8.*
7–8. Nis seo orþung þe wé ut blawaþ and in ateoð oþþe ure sawul, ác ís seo lyft þe ealle lichamlice þing on lybbað, butan fixum anum þe on flodum lybbað. *Lives of Saints, i. 22.*

9. Nis nân licha*m*lic ðing þe næbbe ða feower gesceafta
him mid · þæt is lyft · 7 fyr · eorðe · 7 wæt*er* ;

10. On ælcu*m* lichaman sind þas feower ðing · nimm ænne
sticcan · 7 gnid to sumu*m* ðince · hit hatað þærrihte of ðam
fyre þe him on lutað ;

11. Forbærn ðone oðerne ende · þonn*e* gǣð se wæta ût æt
ðam oðru*m* ende mid ðam smice ;

12. Swa eac ure lichaman habbað ægðer gehǣtan · gewǣ-
tan · eorðan · 7 lyft ;

13. Seo lyft þe we embe sprecað astihð up fornean oð þone
mônan · 7 aberð ealle wolcnu · 7 stôrmas ;

14. Seo lyft ðonn*e* heo astyred bið · is wind ;

15. Se wind hæfð mislice naman on bocu*m* ;

16. Ðanon ðe hê blæwð · him bið nama gesett ;

17. Feower heafodwindas sind · se fyrmesta is east*er*ne
wind · Subsolan*us* gehaten · forðan ðe hê blæwð fra*m* ðære |

f. 251 *r*,   sunnan upsprincge · 7 is swiðe gemetegod ;

18. Se oðer heafodwind is suðerne · Auster · gehaten · se astyrað wolcnu · 7 lîgettu · 7 mislice cwyld blæwð geond þas eorðan ;

---

9. A. Nys   A. þingc   AF. mid him   A. ys   A. lyft 7
desunt   A. fyr · 7 eorþe
10. C. O *for* On   A. synt, BDF. synd   C. þa *for* þas
ACD. Nim, B. Nime, F. nim   A. sumon   A. þingon, BCF.
þinge   A. hyt   F. sone *for* þærrihte
11. A. þæne   F. ænde   A. þænne   B. wætu, F. wæte
F. ðam *deest*
13. A. lyfst   BD. ymbe   A. upp, F. up *deest*   AC.
forneah   A. oþ þæne, C. oð ðene, F. oððe 'þone'   A.
abyrþ, BDF. abyrð   ABCDF. wolcna   F. torras *for* stormas
14. C. astired   A. byþ ys, B. is byð, F. bið *deest*
15. AF. Se wind *desunt*   A. hæfð *deest*   BC. mistlice
16. F. Ðonon   D. ðe *deest*   A. hym   A. bið *deest*, B. byð
CF. geset
17. A. syndon, BDF. synd   F. fyrmysta   A. fyrmesta
wind ys   F. ys   A. wind *deest*   C. gehatæn   ADF.
forþam, C. feoþæn   A. for *for* fram   ABCD. upspringe, F.
upgange *for* upsprincge   AB. ys, F. is *deest*   B. swyðe
18. F. ys   C. suþærne, F. suðþerne   A. astyreþ, B. astyreð,
C. astireð   ABDF. ligettas, C. ligetta   BC. mistlice   F.
cwild   F. blædas *for* blæwð

9–11. Quae (*sc.* elementa) tamen quadam naturae propin-
quitate sibimet ita commiscentur, ut terra quidem arida et
frigida, frigidae aquae: aqua uero frigida et humida, humido
aeri: porro aer humidus et calidus, calido igni: ignis quoque
calidus et aridus, terrae societur aridae. Unde et ignem in
terris, et in aere nubila terrenaque corpora uidemus. *DNR. iv.*

14–16. Uentus est aer commotus et agitatus ... Pro diuersis
autem partibus coeli nomina diuersa sortitur. *DNR. xxvi.*
17–20. Uentorum quatuor cardinales sunt, quorum primus
Septentrio, qui et Aparctias dicitur, flat rectus ab axe, faciens
frigora et nubes ... Secundus cardinalis Subsolanus, qui et
Apeliotes, ab ortu intonans solis, temperatus ... Tertius
cardinalis Auster, qui et Notus, humidus calidus, atque ful-
mineus ... Quartus cardinalis Zephyrus, qui et Fauonius,
hyemem resoluens, floresque producens. *DNR. xxvii.*

14–16. *Cp. Isidore, DNR. xxxvi. 1.*

19. Se ðridda heafodwind hatte zephirus · on greciscum gereorde · 7 on ledenum fabonius · se blæwð westan¹ · 7 ðurh his blæd acuciað ealle eorðlice blæda · 7 blowað · 7 se wind towyrpð 7 ðawað ælcne winter;

20. Se feorða heafodwind hâtte Septemtrio · se blæwð norðan ceald · 7 snawlic · 7 wyrcð drie wolcnu;

21. Þas feower heafodwindas habbað betwux him on ymbhwyrfte oðre eahta windas · æfre betwux þam heafodwindum twegen windas;

22. Þæra naman · 7 blawunge · we mihton secgan · gif hit ne ðuhte æðryt to awritenne;

23. Is swa ðeah hwæðere ân ðæra eahta winda · Aquilo gehâten · se blæwð norðan 7 eastan · healic · 7 ceald · 7 swiðe drie · se is gehâten oðrum naman boreas · 7 ealne ðone cwyld · þe se suðerna wînd auster acenð · ealne he todræfð · 7 afligð;

24. Us ðincð to menigfeald þæt we swiðor embe ðis sprecon;

¹ G. westên

[XI.] DE PLUUIA.

1. Renas cumað of ðære lyfte · þurh Godes mihte;

19. A. þrydda, F. þridde   C. zewirus, F. zepsirus   AC· ledenum bocum fabonius   F. phabonius   C. þe   B. blæð for blæd, D. blæda   D. cuciað   A. blæde, BC. blædu   C. þe   F. towirp   C. þewað
20. A. septentrion   C. þe   B. swawlic, F. snawig   ABCD. drige   C. wolcnum
21. B. betweox   C. heom   C. on ymbhwyrfte desunt   F. embhwyrfte   A. ehta, F. eahte
22. C. þara, F. þare   A. blawungæ, F. blawunga   C. mihtan, F. myhton   A. secgean, B. seggan, C. segcgan   ABF. gyf   A. æþrytt, F. æðrit   B. writenne, F. arwritenne
23. A. ðeah deest, F. þeahð   A. hwæþre, F. hwæðre, B. hwaðere, C. hwæðeræ   C. þære, F. þara   C. se—eastan desunt   F. easten   ABCD. drige   A. ys   C. odrum C. borreas, F. boseas   F. eallne   A. þæne   B. suðrena, C. suðerne   B. acænð   F. hæ
24. CD. the whole paragraph deest   F. þingð   AB. mænigfeald, F. mænifeald   ABF. ymbe   A. sprecan
XI. AC. title deest
1. F. myhte

21–22. *DNR. xxvii passim.*

23. A sinistris (*sc.* Septentrionis) Aquilo, qui et Boreas, nubes constringens. *DNR. xxvii.*

1–2, 7–9. Nubes coacto guttatim aere conglobantur, qui naturali leuitate uapores aquarum de terra marique sustollens, quamdiu in minutissimis stillis consistunt, sua ui suspendit in altum, qua uel solis igne decocti uel aeris itinere mutati dulcescant. *DNR. xxxii.*—Imbres ex nubium concreti guttulis, dum in maiores stillas coeunt, aeris amplius non ferente natura, nunc uento impellente, nunc sole dissoluente pluraliter ad terras dilabuntur. *DNR. xxxiii.*

2. Seo lyft liccað · 7 atihð þone wætan of ealre eorðan · 7 of ðære sæ · 7 gegaderað to scurum · 7 þonne heo mare aberan ne mæg · þonne fealð hit adune to rene alysed 7 toworpen · hwilon þurh windes blædum · hwilon ðurh ðære sunnan hætan;

3. We rædað on ðære bec þe is gehâten Liber regum · þæt se wîtega Helias abæde æt Gode · for ðæs folces ðwyrnyssum · þæt nân rên ne com ofer eorðan · feorðan healfan geare;

4. Ða bæd se wîtega eft æt Gode þæt he his folce miltsian sceolde · 7 him renas 7 eorðlice wæstmas forgifan;

5. Ða astâh he up on anre dune · 7 gebigedum cneowum gebæd for ðam folce · 7 het his cnapan ða hwile behealdan to ðære sæ · gif hê aht gesâwe;

6. Þa æt nextan cwæð se cnapa þæt hê gesawe arisan of ðære sæ · an lytel wolcn · 7 ðærrihte asweartode seo heofen · 7 wolcnu arison · 7 se wind bleow · 7 wearð micel rên geworden;

7. Hit is swa swa we ær sædon · þæt seo lyft atihð up of ðære eorðan · 7 of ðære sæ ealne ðone wætan · þe bið to renum awend;

---

2. B. atyhð, F. astihð   A. þæne, C. þone   F. eallre   A. gegadriað   A. þænne   B. heo *deest*   F. mara   A. þænne   A. hyt   B. windas   AC. blæde, D. blæd, F. blæda   F. þare
3. A. ræddaþ, F. ræðað   A. ys   A. witiga   BF. elias ABCDF. abæd   AC. þwyrnessum, F. þwyrnissum   F. þæta B. com *for* ne com, D. ne come   A. ofor   F. feorðan *deest*
4. AD. abæd   F. æft   D. mildsian, F. myltsian   A. eorðmæstmas, C. eorðwæstmas   A. forgeafe, BF. forgyfan
F *ends here*
5. AD. upp   D. 7—folce *desunt*   B. þæt folc   A. het *deest*   B. is *for* his   A. beheoldon   B. gyf   A. hig *for* he   A. gesawon
6. B. nyxtan   C. cwæð   B. of þære sæ · arisan   A. on *for* of   A. wolc   D. wolcn · 7 wind bleow · 7 asweartode *for* wolcn · 7 ðærrihte asweartode   C. heofon   A. wolcna   C. mycel
7. A. swa *for* swa swa   C. sædan   A. heo *for* seo   B. atyhð   A. upp   A. þæne, C. þonne   A. byþ, B. byð   A. rene   A. gewend

3. Et dixit Elias Thesbites de habitatoribus Galaad ad Achab: Uiuit Dominus Deus Israel, in cuius conspectu sto, si erit annis his ros et pluuia, nisi iuxta oris mei uerba. *III. Regum xvii. 1.*

5–6. . . . Elias autem ascendit in uerticem Carmeli, et pronus in terram posuit faciem suam inter genua sua, et dixit ad puerum suum: Ascende, et prospice contra mare. Qui cum asᴄendisset, et contemplatus esset, ait: Non est quidquam. Et rursum ait illi: Reuertere septem uicibus. In septima autem uice, ecce, nubecula parua quasi uestigium hominis ascendebat de mari. Qui ait: Ascende, et dic Achab: Iunge currum tuum et descende, ne occupet te pluuia. Cumque se uerteret huc atque illuc, ecce coeli contenebrati sunt, et nubes et uentus, et facta est pluuia grandis. Ascendens itaque Achab abiit in Iezrahel. *III. Regum xviii. 42–45.*

---

2. Þære lyfte gecynd is þæt heo tehð to þa renas of þæm sealtan sǽ, 7 þurh hire mægen heo hig fersc sendeð to eorðan. *Martyrologium, p. 40.*

3. On þam dagum wæs helias halig godes witega; se abæd æt gode þæt he ðam yfelan cynincge rén-scuras oftuge, for his reðnysse. Eode þa forð feorðe healfgear butan rén-scurum and reocendum deawe. *Lives of Saints, xviii. 53–57.*

5–6. He astáh ða ardlice up to anre dune, and gebigedum cneowum bæd ðone ælmihtigan god þæt ɲe renas forgeafe eorð-bugiendum, and het his cnapan ða hwile hawian to ðære sǽ, gif ænig mist arise of ðam mycclum brymme; þa gecyrde se cnapa seofon siðum him to, and on ðam seofoðan cyrre sæde ðam witegan, þæt an gehwæde wolcn of ðære widgillan sǽ efne þa upp astige mid þære unscæðþigan lyfte. Efne ða arás se wínd, and ða wolcnu sweartodon, and com ormæte scúr of ðære lyfte. *Lives of Saints, xviii. 142–52.*

8. Ðære lyfte gecynd is þæt heo sicð ælcne wǽtan upp to hîre;

9. Ðis mæg sceawian se ðe wile · hû se wǽta gæð upp · swilce mid smîce · oððe mid miste · 7 gif hit sealt bið of ðære sǽ · hit bið [þurh]¹ þære sunnan hǽtan · 7 ðurh ðære lyfte brâdnysse to ferscum wǽtan awend;

10. Soðlice Godes miht gefadað ealle gewederu · se ðe ealle ðing buton earfoðnysse gediht;

11. Hê nære na ælmihtig · gif him ǽnig gefadung earfoðe wǽre;

12. His nama is om*nipotens* · þæt is ælmihtig · forðan ðe hê mæg eal þæt hê wile · 7 his miht nâwar ne swincð;

[XII.] DE GRANDINE.

Hâgol cymð of ðam rêndropu*m* · þonn*e* hî beoð gefrorene f. 261 *v*. ûpp | on þære lyfte · 7 swa siððan feallað;

[XIII.] DE NIUE.

Snâw cymð of ðam ðynnu*m* wǽtan · þe bið uppatogen mid þære lyfte · 7 bið gefroren ær ðan ðe hê to dropu*m* geurnen sy · 7 swa sǽmtincges fylð;

¹ G. *deest*

---

8. A. ys   AB. sycð   ABCD. up   ABCD. hyre
9. AD. wyle   ABCD. up   B. swylce, C. swilce *deest*   A. swilce he mid smice ga   AB. gyf   C. scealt   AB. byð
A. hyt   A. byþ, B. byð   G. þurh *deest*   A. sunnhætan *for* sunnan hætan   A. wæterum *for* wætan
10. A. gewydera, D. gewyderu   B. gediht buton earfoðnysse
AC. butan   D. earfoðnyssum
11. A. næfre *for* nære   B. gyf
12. A. hys   A. ys   AB. ys   AD. forþam   ABCD eall
D. wyle   A. hys   ABCD. nahwar   A. geswicþ

XII. AC. *title deest*
A. hi *deest*   C. byð   BCD. up   A. of *for* on   A. syþþon, BCD. syððan

XIII. AC. *title deest*
A. ðynnum *deest*   AB. byð   ABCD. up-   AB. byð   AD. þam   D. gerunnen   A. sæmtiges, B. semtinges, CD. sæmtinges

9. In aere nubila terrenaque corpora uidemus. *DNR. iv.*

XII. Grandinis lapilli ex stillis pluuiae, frigoris et uenti uigore conglaciati, in aere coagulantur, sed citius niue soluuntur, et interdiu saepius quam noctu decidunt. *DNR. xxxiv.*
XIII. Niues aquarum uapore, necdum densato in guttas, sed gelu praeripiente formantur, quas in alto mari non cadere perhibent. *DNR. xxxvi.*

10–12. *Cp. v. 6 above.*—Forðanðe his nama is omnipotens Deus; ðæt is on Englisc: ælmihtig God. His willa is weorc 7 he werig ne byð, 7 his mycele miht ne mæg nahwar swincan. *Hexameron, 174–7.*— Deus omnipotens, ðæt is God ælmihtig. *Ælfric's Grammar (ed. Zupitza, p. 196).*

[XIV.] DE TONITRU.

1. Þunor cymð of hætan · 7 of wǽtan · seo lyft tyhð þone
wætan to hire neoðan · 7 ða hætan ufan · 7 ðonne hî gegaderode beoð · seo hǽte · 7 se wǽta · binnon þære lyfte · þonne
winnað hî him betwynan mid egeslicum swege · 7 þæt fyr
aberst ût ðurh ligette · 7 derað wæstmum · gif hê mare bið
þonne se wǽta · gif se wæta bið mare ðonne þæt fyr · þonne
fremað hit · swa hattre sumor swa mare ðunor · 7 liget on
geare;

2. Soðlice ða þuneras þe Iohannes ne moste awritan on
apocalipsin · sind gastlice to understandenne · 7 hî naht ne
belimpað to þam ðunere ðe on ðissere lyfte oft egeslice
brastlað · se bið hlûd for þære lyfte bradnysse · 7 frecenful for
þæs fyres sceotungum;

3. Sy ðeos gesetnys ðus her geendod:— EXPLICIT HEC
BREUITAS DE TEMPORIBUS.

XIV. AC. *title deest* B. DE TONITRUA
1. A. ætan *for* hætan   A. lift   A. tihþ, CD. tihð   A. þæne
AB. hyre   AC. wætan *for* hætan   BCD. ufon   A. þænne
A. hig   A. togædere *for* gegaderode, C. gegaderede   C. hætu
A. binnan   A. þæræ   A. heom   BCD. betweonan   ABD.
abyrst   B. ligett, D. liget   B. gyf   D. hit *for* he   B.
byð   A. wæta · 7 gif   B. gyf   A. byþ, B. byð   A.
þænne   A. þænne   A. hyt · 7 swa   BCD. Swa   ABD.
hattra, C. hatra   AB. mara   B. sumor *for* ðunor
2. A. þunras   A. syndon, BD. synd   A. gaslice   A.
þunre   B. þyssere   A. lifte   B. braslað   A. byþ, B. byð
A. lifte   A. bradnesse   AC. 7 *deest*   ABD. frecenfull   C.
sceotugum
3. C. beo *for* Sy   A. gesetednys, C. gesetnyss
The Latin Explicit *is omitted in* ABCD. B *has instead*: god helpe
minum handum.

XIV. 1. Quidam dicunt, dum aer in se uaporaliter aquam de imis, et ignem caumaliter de superioribus trahat, ipsis confligentibus horrisonos tonitruorum crepitus gigni: et si ignis uicerit, obesse fructibus: si aqua, prodesse. *DNR. xxix.*

XIV. 2. Et (Angelus) clamauit uoce magna, quemadmodum cum leo rugit. Et cum clamasset, locuta sunt septem tonitrua uoces suas. Et cum locuta fuissent septem tonitrua uoces suas, ego scripturus eram; et audiui uocem de coelo dicentem mihi: Signa, quae locuta sunt septem tonitrua, et noli ea scribere. *Apocalypsis x. 3–4.*

NOTES

CHAPTER I

1. For the prefatory paragraph cp. Introduction, p. xlviii, and for the first paragraph cp. Introduction, p. liii.

4. Ælfric seems to have added the description *līchamlic* (of the sky) on his own authority. At least I know of no source for this in Bede or Isidore. Bede, *DNR.* v, expressly states that the sky is *subtilis igneaeque naturae rotundumque.* What *subtilis* means in this connexion is made clear by certain inscriptions on an illumination of Cod. lat. Monac. 13002 (reproduced by F. Saxl, *Verzeichnis astrologischer und mythologischer illustrierter Handschriften des lateinischen Mittelalters. II. Die Handschriften der National-Bibliothek in Wien,* Heidelberg, 1927, p. 42) which may be tabulated as follows:

Ignis—subtilis, mobil[is], acutus
Aqua—corpul[enta], mobilis, obtusa
Aer—subtilis, mobilis, obtusus
Terra—corpul[enta], immobilis, obtusa.

Bede says, then, that the firmament is not corporeal, whereas Ælfric asserts that it is *līchamlic,* or material. Isidore, *De Natura Rerum,* xii. 4, says: *Coelum autem ab Oriente ad Occidentem, semel in die et nocte uerti sapientes existimant. Hoc autem rotundum, uolubile, atque ardens esse dixerunt.* That the sky is of fiery nature, and not solid, is again affirmed in Ælfric's version of Alcuin's *Interrogationes,* no. xx (ed. Tessmann, p. 26): *'Hwilces gecyndes is sêo heofon?'—'Fŷrenes gecyndes and sinewealt and symle turnigende.'* Isidore, Bede, and Alcuin agree, then, that the firmament is round, fiery, and revolving. Ælfric omits the description 'fiery' and says instead that it is *gesewenlic, līchamlic,* and (in § 6) *ansund.* Clark Hall in his third edition renders the latter word by 'solid', referring to a passage in Byrhtferð which, however, is exactly our sentence taken over literally by Byrhtferð from Ælfric.

It might be possible to argue that Ælfric's *gesewenlic* is a misunderstanding of Bede's *subtilis* ('thin' > 'transparent' > 'visible'). It must also be remembered that fire was considered an element and that, therefore, it was thought of as a kind of body or matter: *Ignis quoque materialiter accensus* (Bede, *DNR.* iv). So even if Ælfric tacitly agreed with his sources as to the fiery nature of the firmament he might have rejected Bede's description of it as *subtilis.* But it is much more likely that Ælfric meant fully what he said. The early portions (§§ 1–16) of this chapter are based more on the Book of Genesis itself than on any second-hand account. He himself, in his *Genesis,* translates *firmamentum* by *fæstnys,* and here again he adheres to the literal meaning of the word, disregarding later cosmic speculations. His explanation why the (material) sky was visible and yet could never be seen he found in the common observation of its great distance from the earth. Cp. *Interrogationes*, xxi: *micel swêg gǽð . . . of þâm scînendan rodore, þêah þe wê for þâm mycclan fyrlene hit gefrêdan ne magon.* It is remarkable, though, that MS. A replaces *hī nǽfre gesēon* by *hyne ēac gesēon*.

5. The comparison of the firmament with a mill-wheel is rather inept as it seems to deny the spherical shape of the sky, of which Ælfric really was quite convinced.

7. Cockayne prints *ungesegenlice*, which is supported by MS. B only, and translates 'unspeakable' instead of 'invisible' = *inaccessibile conspectibus* in Bede. Cp. Bülbring, *Altenglisches Elementarbuch*, Heidelberg, 1902, § 490*e*; R. Girvan, *Angelsaksisch Handboek*, Haarlem, 1931, p. 174, § 217; W. Streitberg, *Urgermanische Grammatik*, Heidelberg, 1896, p. 122 sq. *Ungesewenlic* is used in a similar context in *Homilies*, ii. 270; *Lives of Saints*, i. 51 and xvii. 110. Ælfric's word for 'ineffable' is *unasecgendlic*: *Homilies*, i. 286 and 446; ii. 608; *Lives*, i. 33.—Bede devotes chapter vii of *DNR.* to a discussion of the 'upper heavens'. The idea of the upper heavens (usually seven in number) was evolved from the observation that the seven planets had their own course and did not revolve with the firmament as did the 'fixed' stars. It was then assumed that each of these planets was fastened to a sphere of its own, and finally, as happened so often, the astronomical observation was given a theological interpretation: the seven upper heavens were considered the dwelling of seven hosts of angels. Cp. Thorndike, *A History of Magic and Experimental Science*, New York, 1929, p. 487 sq., and my article in *Anglia*, lviii. 294, note 1.

7–9. We have here a pretty example of Ælfric's method of work. Bede's exposition seemed too recondite; he therefore turned to Isidore and found what pleased him better, theology rather than science. But he did not simply translate Isidore but looked up Corinthians and added a translation of the words *et audiuit arcana uerba, quae non licet homini loqui.* When he treated the same matter again in his *Hexameron* he also quoted and translated in full the passage from Psalm 148 to which he refers but briefly in *De Temporibus.*

11. The insistence that the first three days were without the sun goes back to early Christian exegesis, which had to fight the worship of the sun as a deity. Cp. Thorndike, i. 492. Bede, *DTR.* v and *Comm. in Lib. Gen.* (Giles vii. 7 sqq.) discusses at length the nature of primeval light before the creation of the sun. The question would seem to have still been a favourite topic of theological debate in Russia at the time of Dostoevsky. See *The Brothers Karamazov*, iii. vi and v. vi (translated by Constance Garnett, New York, 1915, pp. 128 and 281).

14–15. The scribe of MS. D corrects Ælfric; justly, as reference to Genesis will show.

24. Cp. Tupper, 'Anglo-Saxon Dægmæl', *PMLA*. x (1895), pp. 117–26.

25–30. Ælfric avoids the difficult question of the relative sizes of sun, moon, and earth. Cp. iii. 17 below, F. Boll in Hoops's *Reallexikon der germanischen Altertumskunde*, i. 135, and Thorndike, i. 521.

31. Berthold von Regensburg's homily *Sælic sint die reines herzen sint* (ed. F. Pfeiffer, Wien, 1862, i. 388–407) parallels *De Temporibus Anni* in a number of places. The following passages correspond: B(erthold), 390, 4–13 = *DTA*. i. 31, 38–40 (the stars borrow their light from the sun). B. 391, 17–21 = *DTA*. i. 38–9 (quotation from John, i. 9). B. 391, 27–8 = *DTA*. i. 32 (the sun outshines all the stars). B. 392, 21–2 = *DTA*. vi. 9 (the earth like a ball). B. 392, 22–9 = *DTA*. i. 5 (the firmament encircles the earth and revolves like a wheel). B. 392, 29–30 = *DTA*. ix. 14 (the learned and the unlearned; cp. Reum, op. cit., p. 493; Crawford, *Exameron Anglice*, p. 152; *Anglia*, lviii. 304, note 3). B. 393, 12–13 = *DTA*. v. 6 (the earth suspended in mid-air, supported by God). B. 393, 18–26 = *DTA*. iii. 3–5 (night caused by the earth's shadow). B. 400, 28–30 = *DTA*. iii. 7 and iv. 26 (the moon the lowest of the stars; cp. also Berthold, i. 53, 24–5 and ii. 235, 20). B. 400, 31–401, 14 = *DTA*. iii. 15–16 (lunar eclipse). B. 400, 33–6 = *DTA*. iii. 17 (relative size of earth and moon).

33–40. Cp. F. Boll, *Sternglaube und Sterndeutung*, 3rd edition by W. Gundel (1926), p. 31; *Anglia*, lviii. 294, note 1. Bede, *DTR*. lxiv, likens the moon to Christ, but also to the Church: *Luna eo suae lucis incremento, quod exiens a sole ad nostros reuoluit obtutus, domini saluatoris in carne, usque ad tempora passionis doctrinam uirtutesque significat, eo autem quod ad solem rediens paulatim ad inuisibilem nobis coeli faciem recolligit, resurrectionis illius ac posterioris gloriae miracula demonstrat* (Giles vi. 267).—*Septem quoque dies lunae . . . uniuersitatem ecclesiae, quae per totum mundum paschalibus est redempta mysteriis, aperte denuntiant* (Giles, vi. 268).

38–40. The quotation from John has an echo in a twelfth-century homily on the assumption of the Virgin. I quote from this homily not only what parallels the present passage but also the reference to the *sǣsterre*, which throws some light on chapter ix. 8 below: *Hire is to name Maria quod est interpretatum stella maris Đat is on englis sæ sterre. Đan þe safarinde men seð þe sa sterre. hie wuten sone wuderward hie sullen weie holden. for þat þe storres liht is hem god latðæu . . . And alse þe sa storre shat of hire þe liht. þe lihteð sa farinde men. alse þis edie maiden seinte Marie. of hire holie licame shedeð þat soðe liht. þe lihteð alle brihte þinges on eorðe. and ec on heuene. alse Seint Iohannes saið on his godspel. Erat lux uera que. illuminat omnem hominem uenientem in hunc mundum. He is þat soðe liht. þe lihted alle men. þe on þis woreld cumeð. and aleomed ben. And for þis leome is þat holie maiden cleped sa sterre. R*. Morris, *Old English Homilies*, Second Series, London, 1873 (E.E.T.S. 53), p. 161.

CHAPTER II

4. Most Anglo-Saxon calendars note the *primus dies saeculi* on
March 18th. Cp. R. T. Hampson, *Medii Ævi Kalendarium*, London,
1841 (especially ii. 119 sq.); F. Piper, *Die Kalendarien und Martyrologien der Angelsachsen*, Berlin, 1862, p. 86 sq; F. Wormald, *English
Kalendars before A.D. 1100*, London, 1934; H. Henel, *Studien zum
altenglischen Computus*, Leipzig, 1934, p. 76. Cp. also the OE.
Menology, ed. Imelmann, Berlin, 1902, ll. 44-7.

CHAPTER III

2. This paragraph should be read in connexion with iii. 14. In the
latter place Ælfric states what was a common belief in the Middle
Ages, viz. that the empty spaces above the earth's atmosphere and
below the ceiling of the firmament were continuously lit by the sun
and the stars. The moon was supposed to run just about on the border
of the air and the empty ether. We know, of course, that our atmosphere does not extend nearly as far as the moon, and that above the
atmosphere, although the sun and the stars can be seen, light
is not diffused as it is on earth, because there is no matter to
reflect the sun's rays. This misunderstanding about the physical
nature of the universe seems to have given rise to the idea, expressed
in iii. 2, that in the heavens, which were supposed to be located in
concentric circles above or beyond the firmament, there was eternal
light; a conception which also had its theological significance and
which agreed with the ancient symbolization of God, or Christ, as the
sun. Cp. the note on i. 33 above. The idea persisted through the centuries, as did the parallel concept that in heaven there is eternal spring-
time. The latter is found as the motto of a poem by Simon Dach
(1605-59): *Perpetui coelum tempora veris habet* (M. Sommerfeld,
Deutsche Barocklyrik, Berlin, 1934, p. 71). Just as German *Himmel*
combines the meanings of 'sky' and 'heaven', so Ælfric uses the word
heofon now in the sense of *rodor* or *firmamentum*, 'sky' (cp. v. 1–3),
now in the sense of 'heaven'.

7. Ælfric's translation, or rather condensation, of Bede's exposition is not very clear in this place, so that Cockayne, p. 243, mis-
translated the passage. John Lingard, *The History and Antiquities
of the Anglo-Saxon Church* (Rolls Series), London, 1845, ii. 175, says:
'Beda explains with sufficient accuracy the causes of the solar and
lunar eclipses, and observes, that their recurrence at each conjunction and opposition is prevented by the obliquity of the moon's orbit.'
Instead of writing *for ðām brādan circule* Ælfric should have said *for
ðǣre brādnysse ðæs circules*, because it is not the wide *circle* zodiac,
but the *width* (about 8 degrees from the ecliptic on either side) of that
circle, which allows the moon to pass the sun without obscuring it, and
to pass around the earth without being obscured by it: *Siquidem
Signifer idem CCC quidem LXV partibus et quadrante per coeli*

*ambitum longus, sed XII est partibus latus; harum duas tantum medias
sol, luna omnes peruagari consueuit* (Bede, *DTR*. xxvi). A simpler way
of saying this would be that the ecliptic and the moon's orbit are
not identical. Byrhtferð has a diagram (reproduced by Crawford
opposite p. 164) showing the phases of the moon, but from it one
would gain the impression that the eclipses are a regular event at each
conjunction and opposition, even though he marked the 'width' of
the zodiac in a naïve fashion.

9. A *prica* is the fourth or fifth part of an hour. Cp. Bede, *DTR*. iii:
*Recipit autem hora IV punctos . . . et in quibusdam lunae computis V
punctos.* It is here the fifth part of an hour, as is apparent from the
parallel passage, *DTR*. xxiv: *Et ideo si nosse uis luna secunda quot
horas luceat, multiplica per quatuor duo, fiunt octo, partire per quinque,
quia quinque puncti horam faciunt, quinquies asse quinquis et remanent
tres, horam ergo et tres punctos lucet luna secunda.* It is remarkable
that Byrhtferð (ed. Crawford, p. 76), or perhaps the scribe, changes
three times the word *quattuor* (*puncti*) into *quinque* when he deals
with the *saltus lunae.* He must have had in mind the value of the
prica or *punctus* as relating to the moon's age, although here he was
concerned with solar hours and their divisions. Cp. Byrhtferð, 112.
28: *Feower puncti, þæt synt prican, wyrcað ane tid on pære sunnan
ryne.* Again when he translates (Crawford, 160–2) Bede's chapter
Quot Horis Luna Luceat (*DTR*. xxiv) Byrhtferð says: *Anre nihte eald
monan scynð feower prican, ac we wyllað secgan to soðe fif prican æfter
Bedan gesetnysse.* Now Bede says no such thing, and Byrhtferð disre-
gards his own remark in the calculation which follows. It is clear that
Byrhtferð himself was not very certain of the distinction involved,
viz. that a *prica* was the fourth part of an hour in ordinary usage, but
that it was accounted the fifth part of an hour in relation to the moon's
progress through the zodiacal signs (cp. *DTR*. xvii).—Cp. Bosworth-
Toller, s.v.; Dietrich, *Niedners Zeitschrift*, xxvi. 165; Tupper, *PMLA*.
x. 131; Henel, *Studien*, p. 66. Also the note on viii. 13–15 below.

14. See the note above on iii. 2. Ælfric's translation of *spatia ultra
lunam* by *þæt fæc bufon ðære lyfte* is justified by the ancient conception that the moon runs her course just about on the border of the
air and the empty ether. Cp. x. 4 and the passage from Bede's
Commentarii in Lib. Gen. which is quoted as a parallel to this paragraph. On the other hand, the ether itself was considered an element
(the fifth) in Aristotelian philosophy and by some later schools. Cp.
Franz Dornseiff, *Das Alphabet in Mystik und Magie*, 2nd edition,
Leipzig and Berlin, 1925, p. 83.

17. 'Respecting its (*sc.* the moon's) magnitude, the Saxons followed
two opposite opinions. Some, on the authority of Pliny, maintained
that it was larger; others, with greater truth, conceived that it was
smaller, than the earth.' John Lingard, *The History and Antiquities
of the Anglo-Saxon Church* (Rolls Series), London, 1845, ii. 175.
Bede, *DNR*. xxii, still adheres to Pliny whilst Ælfric deftly side-steps

the issue. Cp. Thorndike i. 488. For the views and calculations of the ancient Greeks see I. L. Heiberg, *Geschichte der Mathematik und Naturwissenschaften im Altertum*, München, 1925, pp. 53, 56. 18–25. The list given in the Parallels from MS. Additional 32246 is printed (incorrectly) in Wright-Wülker, *Anglo-Saxon and Old English Vocabularies*, London, 1884, col. 175, as part of what is there called a 'Supplement to Ælfric's Vocabulary'. Neither the Vocabulary itself nor the Supplement are Ælfric's, as was shown by M. Förster, *Anglia*, xli. 94 sqq. The list itself is quite corrupt in the MS. The following is an attempt to restore its original order:

(1) Crepusculum: *æfenglōma*[1] uel *twēonelēoht* uel *peorcung*.
(2) Uesperum uel Serum: *æfen*.[2]
(3) Conticinium: *swīgtīma*.[3]
(4) Intempestum: *midniht*.
(5) Gallicinium: *hancrēd*.
(6) Matutinum: *ūhtentīd*.
(7) Diluculum: *dægrēd*.
(8) Aurora: *dægrima*.[4]

For an exhaustive discussion of the divisions of night cp. Tupper, *PMLA*. x. 126–8, 135–87.

26. Ælfric omitted a discussion of weeks and months partly because Bede's treatment of them is very full (although not particularly difficult, as Ælfric claims), but also because he was afraid he might give further impetus to the many superstitious practices that were based on the days of the week and their names. Cp. *Anglia*, lviii. 305 and 317.

27–29. It is impossible to note all the places in which the criteria for the reckoning of Easter are given, and equally impossible to say with certainty which of them may have been Ælfric's source. The following passage from the computus of the *Missal of Robert of Jumièges* (ed. Wilson, p. 39) agrees fairly closely with Ælfric's exposition: *Sanctae memoriae Theophilus Alexandrinus episcopus datis epistolis ad Theodosium imperatorem in quo adnuntians ab VIII. id. mar. usque in diem non. apr. dies scilicet XXIX qualiscumque luna nata fuerit in quolibet medio spatio perhibet facere initium primi mensis. XIIII*[a] *uero luna a XII*[a] *kal. apr. usque in XV*[am] *kal. mai sollerter inquiri etiam si die sabbatorum inciderit consequenti die dominico. id. est luna XV pascha caelebrare conscripsit et si die dominico luna XIIII eiusdem mensis id est primi mensis euenerit ipsa ebdomada transmissum ad alterum diem dominicum pascha caelebrare sine dubio conscripsit*.[5] The passage is translated by Byrhtferð, p. 138, 7–16. He states the criteria no less than six times, viz. pp. 80, 2–12; 84, 9–17; 136, 12–24; 136, 25–138, 6; 144, 4–21; 168, 12–21. The

[1] *dægred* MS. [2] In the MS. line (2) follows line (3).
[3] *hancred uel Gallicantum* MS.
[4] The last two lines should probably be one.
[5] The Latin is corrupt; see below p. 95, note on vii. 1–8.

Anglo-Saxon computi always have a number of paragraphs setting forth the rules governing the calculation of Easter. Cp. *Studien zum altenglischen Computus*, Parts B and C. In the Old English Menology of MS. Tiberius B.I. (ed. Imelmann, Berlin, 1902) reference is made to the difficulties of the computation of Easter: *Ne magon we þâ tide be getæle healdan dagena rimes, ne drihtnes stige on heofenas up, forþan hê hwearfað âá wisra gewyrdum; ac sceal wintrum frod on circule cræfte findan halge dagas* (ll. 63–8).

CHAPTER IV

3–14. The Old English names of the zodiacal signs are also found in the Calendar of MS. C.C.C.C. 422 and printed by F. Wormald, *English Kalendars before A.D. 1100*, London, 1934, pp. 184–95.

18–21. For a full discussion of the questions relating to the various beginnings of the year cp. Grotefend, *Zeitrechnung des deutschen Mittelalters und der Neuzeit*, vol. i, s.vv. *Annunciationsstil, Circumcisionsstil*; also Tupper, *PMLA*. x. 207–12; Henel, *Studien zum altenglischen Computus*, p. 87 sq.; *Englische Studien*, lxix. 341 sq.— Cp. also *Martyrologium*, ed. Herzfeld, p. 2.

22, 29–30. The zodiac was divided into 360 'parts', and it was supposed that the moon daily proceeded 13 'parts' and the sun only one: *illa* (sc. *luna*) *XIII partes complente sol unam complet* (Bede, *DTR*. xviii). What Ælfric describes as the lunar year is the course of the moon around the zodiac: it takes $27\frac{1}{3}$ days to make this way. The calculation is near enough, although not quite correct: $27\frac{1}{3} \times 13 = 355\frac{1}{3}$ 'parts'. It should, of course, be 360 'parts', which proves that the moon daily does a little more than 13 'parts'. Meanwhile the sun has also moved away from the starting-point, and the moon takes another 2 days and 4 hours to catch up with him. This, the period from one conjunction to the next, Ælfric calls the lunar month and gives to it $29\frac{1}{2}$ days (in actual fact it is 44 minutes and 3 seconds more). Again the reckoning is close enough for practical purposes: in this period the sun moves $29\frac{1}{2}$ 'parts', and the moon $29\frac{1}{2} \times 13$, or $383\frac{1}{2}$ 'parts'. The moon is thus $23\frac{1}{2}$ 'parts' beyond its starting-point. The discrepancy of 6 'parts' (the moon should now be exactly where the sun is) arises because, on the one hand, the moon does more than 13 'parts' per day (as we have said) and, on the other, because the sun does a little less than one 'part' per day. (It takes 365, not 360, days to traverse the 360 'parts' of the zodiac.)

31–33. Cp. F. Rühl in Hoops's *Reallexikon*, iii. 237, and *Studien zum altenglischen Computus*, p. 38 sq.

34–35. The rule which Ælfric states in this place suffered exceptions in the eighth, eleventh, and nineteenth years of the *circulus decennovenalis*. Cp. Byrhtferð, pp. 104, 17–21; 106, 25–7; 110, 16–24; also 154, 23–7 and 156, 24–30. Cp. further Grotefend, i. 128, 163, and *Studien zum altenglischen Computus*, pp. 20 and 25.

35. Instead of the more common word for March, *Hrēðmōnað*,

Ælfric uses regularly *Hlȳda*. *Hlȳda* is found only in Ælfric, in the OE. Menology, and in two short texts. One of these, *De Diebus Malis* (*Englische Studien*, lxix. 336), most probably was influenced by Ælfric's homily *Octabas et Circumcisio Domini*. Cp. *Studien zum altenglischen Computus*, p. 11 sq.

36. Ælfric's word for 'season' is *tīd*, Byrhtferð's *tīma*. Cp. Richard Jordan, *Die Eigentümlichkeiten des anglischen Wortschatzes*, Heidelberg, 1906, p. 92 sq.

CHAPTER V

7–9. It will be noted that in this passage Ælfric is much more explicit than the extract from Bede which I have quoted as his source. He denies, by inference at least, the existence of the celestial oceans which play such a part in the cosmologies of Bede (*DNR*. viii) and Isidore. The following passage from Isidore's *DNR*. xlviii is nearer to Ælfric in some respects. It agrees entirely with the latter's conception of the location of the earth (v. 6) and parallels, although less exactly, his discussion of the sea: *Nunc terrae positionem definiemus, et mare quibus locis interfusum uideatur, ordine exponemus. Terra, ut testatur Hyginus, mundi media regione collocata, omnibus partibus coeli aequali dissidens interuallo centrum obtinet. Oceanus autem regione circumductionis sphaerae profusus, prope totius orbis alluit fines. Itaque et siderum signa occidentia in eum cadere existimantur.* Isidore's distinction between the 'sphere', whose periphery or outer region is covered by the ocean, and the 'orb', the shores of which are washed by it, may not at once seem clear. By *sphaera* he means the globe, by *orbis* the more or less disk-like tract of land then known. He is harking back to the views of the Greek geographers who, although quite convinced of the spherical shape of the earth, spoke of the οἰκουμένη as an island in the ocean (cp. Heiberg, op. cit., p. 84). These views seem to retain, side by side and not fully reconciled, both the Homeric conception and that of later Greek science. The same is true of Bede, *DNR*. xliv. Ælfric, in resolutely rejecting the view that the ocean encircles the earth, is definitely more modern and may, therefore, have had sources other than Bede or Isidore for this passage.

8. This sentence is a reference, conscious or unconscious, to the so-called Legend of Adam's Creation, which correlates the parts of the human body with certain elements of the physical world. The Legend was originally part of the apocryphal Book of Enoch and was known widely in the Middle Ages. Cp. Max Förster, 'Adams Erschaffung und Namengebung', *Archiv für Religionswissenschaft*, xi. 477 sqq., and my article in *Anglia*, lviii. 312 sqq.

CHAPTER VI

1–6. The vexed question regarding the dates of the equinoxes and solstices had really been settled for the English Church by Bede, but

it still troubled the reckoning of Easter in Ælfric's time. Although
Ælfric is here translating almost literally he is not dealing with a dead
issue. This is shown by the confusing variety of the entries in English
calendars of the tenth and eleventh centuries, which will be found tabulated in *Studien zum altenglischen Computus*, p. 75. The list there given
may now be augmented and corrected by reference to F. Wormald's
edition of the *English Kalendars before A.D. 1100*. For the question in
general cp. Grotefend, i. 90 and Piper, *Die Kalendarien und Martyrologien der Angelsachsen*, Berlin, 1862, p. 83.

5. For the translation of *þa gelēaffullan fæderas* see Introduction,
p. xlii. Cp. also *Lives of Saints*, Preface l. 50 sq.: *geleaffulle fæderas
and halige lareowas hit awriton on leden-spræce*; ibid., i. 84: *þa geleaffullan fæderas þe godes lare awriton*; and *Homilies*, ii. 444: *ða
geleaffullan lāreowas, Augustinus, Hieronimus, Gregorius.*—Cockayne
translates *gewisse dægmæl* by 'sure day measurements', Bouterwek
(*Screadunga*, p. 69) by 'signum diei'. Ælfric uses the word *dægmæl*
as the equivalent of Bede's *horologica consideratio*, which seems to
indicate that he had in mind the simple meaning of 'dial'.—The
sun-dial was the chief chronometer of the Anglo-Saxons. Cp. Tupper,
loc. cit., pp. 128–32; Ernst Zinner, *Geschichte der Sternkunde*, Berlin,
1931, p. 346. Where dials were not available they observed the shadow
of the human body. Several Anglo-Saxon calendars give the length of
the shadow (presumably of a 6-foot gnomon) for each month. Cp.
Studien zum altenglischen Computus, pp. 29 and 59; Wormald, op.
cit., *passim*.

7. The (approximate) length of day and night in each month is
noted in practically all Anglo-Saxon calendars. Cp. *Studien*, p. 17;
Wormald, *passim*; also i. 24 and vi. 2 above. Tupper, loc. cit., pp.
117 sqq., has shown that day and night were each divided into twelve
equal parts, so that not only the days but also the hours were longer
in summer than in winter.

9–10. Cp. vi. 20 below.—That the earth is round, 'not like a
shield, but rather like a ball', was already asserted by some Greek
philosophers (Aristotle) as well as by some fathers of the Church,
notably Basil. Cp. Heiberg, op. cit., p. 50; Thorndike, op. cit. i.
480 sq.; Lingard, op. cit. ii. 174, note 2. F. Boll in Hoops's *Reallexikon*, i. 135, claims that since the eighth century A.D. the modern idea
of a sphere-shaped earth had become firmly established. Ælfric,
however, cannot have been fully aware of the implications of the
modern conception, as is shown by his remarks in i. 28 and in
Homilies, i. 286.

11. This passage is an illustration of the well-known fact that
medieval scholars accepted and copied the findings of ancient science
and geography merely as facts. They did not understand the mathematical basis of Greek astronomy and therefore were troubled little
by any inconsistencies which arose through careless or confused tradition. In the present instance Ælfric's, or rather Bede's, information

is fairly accurate. Within a zone 23° 27′ north of the equator the sun casts its shadow south in summer and north in winter, although it is only on the equator itself that the shadow lies each way for exactly six months. Since the southern tip of India touches latitude 8° north one might be tempted to think Ælfric really knew where India was situated. In actual fact, however, he did not have any knowledge of its position or climate: *Wyrd-writeras secgað þæt ðry leodscipas sind gehâtene India. Seo forme India lið to ðæra Silhearwena rice, seo oðer lið to Medas, seo ðridde to ðam micclum garsecge; þeos ðridde India hæfð on anre sidan þeostru, and on oðere ðone grimlican garsecg* (*Homilies*, i. 454). Isidore's *Etymologiae*, xiv. iii and v, could have been consulted for a more accurate description of the geography of India, had Ælfric been at all aware of the inconsistency of the traditions he followed.

12. Here is another example of the 'blindness' of medieval learning. Bede reports that at the (summer) solstice the sun is in the zenith at Syene in Upper Egypt. It was through observation of the sun and its shadow at Syene and at Alexandria, and through measuring the distance between these cities, that Eratosthenes (275–195 B.C.) accomplished the memorable feat of calculating the earth's circumference (Heiberg, op. cit., p. 84). Bede's remark thus misses the point. Not only at Syene, but at any location under the tropic of Cancer the sun is in the zenith on June 21st, and thus the special importance of Syene is historical, not geographic. Ælfric makes matters worse by simplifying his source. He substitutes Alexandria for Syene, but the latitude of the former city is about 31° 40′ north, or more than 8° beyond the zone in which the sun may be observed in the zenith.

14. Ælfric correctly renders Bede's *VIII partes unius horae* by 'a little more than half an hour'; cp. *DTR.* iii: *Recipit autem hora IV punctos, X minuta, XV partes.* Cp. also Tupper, p. 131, and my *Studien*, pp. 65–8.

18. In addition to the remarks in the Introduction, p. xlv, it might be mentioned that Ælfric uses identically the same phrase, *swa swa we gesawon sylfe foroft*, in his homily on St. Swithhun, *Lives of Saints*, xxi. 263. In the latter place he relates how the monks of the Old Minster at Winchester were at first slack in obeying Bishop Æthelweald's commands, but later observed the custom of singing the *Te Deum* whenever a person was healed through the help of St. Swithhun. Ælfric adds: 'as we ourselves have very often seen, and have not seldom sung this hymn with them' (Skeat's translation). J. H. Ott, *Über die Quellen der Heiligenleben in Ælfrics 'Lives of Saints I'*, Halle, 1892, p. 52, claims that Ælfric's lines 223–64 are translated from his sources, Landferth's *Historia translationis et miraculorum* and the *Miracula S. Swithuni*. This is not true, however, of the crucial ll. 262–4, as reference to the *Acta Sanctorum*, July 2nd, pp. 293 and 295, will show. The *Historia* merely says: *quod* (sc. *mandatum Athelwoldi) ita deinceps hactenus observatum audivimus,*

whilst the *Miracula*, which parallels Ælfric's text much more closely
in this place than does the *Historia*, has nothing even vaguely
resembling Ælfric's ll. 262–4. The miracles referred to began only
after the translation of the remains of St. Swithhun from the church-
yard outside to the Old Minster itself. This translation took place
July 15, 971 (cp. Ott, p. 7, White, p. 37, *DNB*. xix. 240), and since
Ælfric's sojourn at Winchester is usually placed in the years 972–87
(cp. White, p. 11) there is little doubt that he really speaks from
personal experience. Thus it would seem that a similar claim in
De Temporibus Anni cannot be dismissed lightly. Miss White says
on p. 35 of her monograph: 'There is found in his writings no trace
of his early home and parentage. It can hardly be doubted that he
was a Wessex boy.' This view may now have to be modified.

19. It is noteworthy that in this place Ælfric corrects his source.
Bede's remarks about the length of days and nights in the polar regions
are rather muddled. He says (*DTR*. xxxi *in fine*) that at the poles
daylight is continuous for six months, and night for the following six
months, and that in the island of Thule (Iceland) there are, at the
time of the summer solstice, a few days without any night. But then
again he confuses the two facts, quoting Pythias of Marseille as saying
that in Thule itself there is continuous day for six months. Yet he
knows (cp. *DNR*. ix *in fine* and *DTR*. xxxiv [Giles, p. 216]) that Iceland is a long way off the pole. Ælfric seems to have recognized the
error. He writes that there is no night in Thule 'for six days' (instead
of Bede's 'for six months'). Even this, however, gives a more
northerly position to Iceland than it actually has. The north coast
of Iceland just touches the Arctic circle, and under that circle there
is only one day of twenty-four hours without any night. Whether
Ælfric made his correction from a real knowledge of the position of
Iceland is, therefore, rather doubtful. Indeed it might be argued that
he merely repeats the words 'six days' which occur a few lines earlier
in both Bede's and his texts, when it is said that Thule is six days'
voyage north of Britain.

25. Ælfric again deviates from his source, but this time with less
justification, when he says that only the north temperate zone is
inhabited. Bede explains that both temperate zones are inhabitable,
but adds *quamuis unam solummodo probare possunt habitatam*.

CHAPTER VII

1–8. The passage quoted from the Missal of Robert of Jumièges is
obviously corrupt in several places. The bissextile day has, of course,
twenty-four hours and not twelve. The beginning of this same Latin
text is also found in MS. Cotton Tiberius B.V, fol. 18*r*, and its end
in MS. Titus D. XXVII, fol. 24*v*. These copies are even less reliable.
The opinion that Joshua's prayer caused the bissextile day is expressed in a text of the Leofric Missal, ed. Warren, p. 21: *Inquiren-
dum est quare dicitur bisexus. Dicitur bisexus propter his kalendas*

nominatas, et ut dii quando opugnauit iosue in terra gabaon orauit ad dominum ut staret sol tribus oris in celo, et per optineret uictoriam. Et ita factum est quasi annis singulis ipsi hore adcrescunt. Et in tres annos ad quartam faciunt unum diem. Et ipse dies dicitur bisexus. . . . This text also requires emendation, but I have repeated it as printed by Warren since the meaning at least is clear. Yet another Latin text on the intercalated day is found in MS. Caligula A. XV, fol. 127*v*. It does not mention Joshua, but adds a sentence in Old English (printed by Förster, *Englische Studien*, lx. 75).

1. The word *prēostas*, 'priests', has a very special meaning in this context. It means practically as much as *rīmcræftige weras,.* 'computists'. In a number of places Byrhtferð contrasts the art of the *bōceras*, 'scholars', with that of the priests, and he makes it clear that the computus is the real study of priests, whereas literary pursuits are not: *Ars grammatica inimica est deo* (Ehrismann, *Geschichte der deutschen Literatur bis zum Ausgang des Mittelalters*, München, 1918, i. 140). I quote the instances in Crawford's translation: 'It appears to me that the "leap" of the priests (i.e. *saltus lunae*) surpasses this literary "leap" [a spondee instead of a dactyl], because it is mysterious, and necessary to know' (Byrhtferð, p. 100, 18–20). 'Grammarians in their art have excellent symbols and signs, like these which I intend to make known to priests. Moreover, it seems to me delightful to make known to grammarians the marks of the priests, lest they should charge the computists with being without scientific signs' (Byrhtferð, p. 182, 22–6). 'Scholars tell us that the hexameter verse must have twenty-four "morae", and priests skilled in the computus [*rīmcræftige prēostas*] say that the day must have twenty-four hours' (Byrhtferð, p. 192, 8–10). 'First of all we will write them [the letters of the alphabet] down together, and then we will make known their divisions in the manner which scholars have and hold [i.e. vowels and consonants apart]; and likewise we will group the letters separately, which devout priests have in their reckoning [i.e. the Roman numerals]' (Byrhtferð, p. 194 sq.). Cp. also Byrhtferð, 94, 12–19. It is apparent, then, that Byrhtferð repeatedly uses the word *prēostas* in the sense of 'computists'. In the passage under consideration Ælfric uses it in the same way. It is his only reference in *De Temporibus Anni* to a source other than Bede and, as was shown in the preceding note, he does here draw upon a Latin text not found in Bede, but occurring in three Anglo-Saxon computi.

3. *Tuwa six* is a mistranslation of 'bissextus' (cp. *Anglia*, xi. 475). It should, of course, be *tuwa se sixta*. Since Ælfric correctly renders 'sextus' by *se sixta* it might be argued that his original cannot have contained the error, and that the passage should be emended against the evidence of all existing MSS.

5. Bede's *aequinoctialis solis locus in signifero circulo* is inexactly rendered by *þæs emnihtes circul*. Cockayne, to put the matter right, added in his translation (pp. 263–5) the words: 'that is, it [the sun]

crosses the equator.' Both here and in iii. 25 *ǣrmergen* seems to refer to sunrise, or six o'clock in the morning, although Tupper, *PMLA*. x. 152, says: '*Ær-morgen* may be regarded as extending from Dawn to Undern (Mid-morning).'

7–9. The Anglo-Saxon calendars regularly note the intercalated day on February 24th. Many of them also add on the foot of the page a Latin note reminding the computist that in a leap-year he must give thirty days (instead of twenty-nine) to the moon of February. There are two versions of this Latin text and their occurrence helps to classify the MSS. in which they are contained. It very probably goes back to the passage in Bede's *DTR*. xli which is quoted above as Ælfric's source. Byrhtferð also translates this passage from Bede (his translation is found in the parallels above), but for good measure he goes on to say the same thing again by translating also the Latin note in the calendars (ed. Crawford, p. 72, 6–13). Cp. *Studien zum altenglischen Computus*, p. 19 and note 57.

CHAPTER VIII

1–2. The passage from Byrhtferð quoted in the Parallels is borrowed from Ælfric. The same matter is also given by Byrhtferð on p. 74, 8–11 where he translates from Bede directly.

5. The quotation from Bede given as the source of this paragraph shows that Ælfric is not right when he says that the dispute about the moon's age is one between the learned and the unlearned. Bede discusses the matter at length in *DTR*. xliii and explains that the computation of the moon's age for purposes of the reckoning of Easter does not necessarily coincide with the actual appearance of the moon in the sky. The incident of the new moon is laid down by Church authority. It is a matter of faith rather than of science: *Ideoque hanc* (sc. *lunam paschalem primam*) *aliter definire nulli fidelium fas est* (Giles, vi. 232). The opinion of the 'learned', as Ælfric calls them, therefore corresponds to the doctrine of the Church, whereas the 'unlearned' are those who would trust observation rather than tradition. By doing this, however, they would spoil the cyclic computation of the Easter moon.

7–11. In this passage Ælfric forbids the taking of auspices from the moon. Following his source (Bede's *DTR*. xxv) he mentions only one type of such prognostication, forecasting the weather by the *position* of the moon in the sky. In his own time, this particular type does not seem to have been known; at least no texts explaining it have come to light in English MSS. of the tenth or eleventh centuries. But in the homily on the Octave of the Lord (the relevant passages are quoted in the Parallels above) Ælfric also mentions that people avoid doing business on Monday, *dies lunae*, (cp. Förster in *Studies in English Philology . . . in Honour of Frederick Klaeber*, pp. 270–7) and that they will not undertake journeys on certain days of the moon. The latter type, superstitious beliefs and forecasts connected with the *phase* of the moon, was very common in Anglo-Saxon England. This is attested by a surprisingly large number of entries, both in Latin and in Old English, in MSS. about coeval with Ælfric, or a little later. There are texts foretelling the health, character, and fortunes of a child born on a given lunar day (Förster, *Archiv für Neuere Sprachen*, cxxix. 16–30). Others foretell the prospects of a man falling ill on a certain lunar day (ibid., pp. 30–6). Others again advise about what to do and what not to undertake on such days (ibid., pp. 37–45). Yet another type tells whether dreams will come true or not by the lunar phase on which they were dreamt (Förster, *Englische Studien*, lx. 58–93). Again there are lists showing on what lunar days a person may be phlebotomized with safety (my article in *Englische Studien*, lxix. 333–5); and, finally, there are texts combining the information of the five types mentioned (Förster, *Archiv für Neuere Sprachen*, cxxi. 32–3). It is an ironical fact that MSS. A and D of *De Temporibus Anni* are among the richest sources for prognostic texts of this nature. In MS. D, *De Temporibus Anni* is immediately preceded by a Latin text of the last-named (i.e. the most comprehensive) type. Some of the others are listed in the Introduction, p. xiii above. One is forced to the conclusion that Ælfric's warning was of little avail, indeed in one case it was hardly even understood (*Englische Studien*, lxix. 342). This is not surprising when one considers that what follows in *De Temporibus Anni* (viii. 12–14) is scarcely less fantastic than the superstitions which Ælfric condemns, or that Bede, *DTR*. xxx, repeats on the authority of Hippocrates advice as to when a man may or may not eat, drink, love, or wash his hands with safety. Cp. *Handbuch des deutschen Aberglaubens*, s.v. *Mond*; Thorndike, i. 492 and 632; Philippson, *Germanisches Heidentum bei den Angelsachsen*, Leipzig, 1929, p. 125; Wulfstan, ed. Napier, p. 104 sq.; Emanuel Svenberg, *De Latinska Lunaria*, Göteborg, 1936, pp. 1–7. The most detailed warning against superstitious practices which Ælfric ever gave is his *Sermo in Laetania Maiore: De Auguriis* (*Lives of Saints*, xvii). Strangely enough, prognostication by the moon is not mentioned in it at all, and soothsaying by the day of the week is referred to only in passing (ll. 92–5).

9. Ælfric uses *onǽlan* or *ātendan* for the 'kindling' of the moon by the sun. The metaphor is borrowed from the Latin *accendere* (cp. *Studien zum altenglischen Computus*, pp. 43 and 58) and is very ancient (cp. H. Usener, *Götternamen*, 2nd edition, Bonn, 1929, p. 288).

12. The word *cepan* is used regularly by Ælfric in speaking of the observation of portents, natural or superstitious: *Ne sceal nan man cepan be dagum on hwilcum dæge he fare, oððe on hwylcum he gecyrre* (*Lives*, xvii. 92 sq.); *We ne sceolan cepan ealles to swyðe be swefnum* (*Lives*, xxi. 403 sq.). The possibilities of forecasting the weather by the colour of sun, moon, and sky are treated at some length in Isidore's *De NaturaRerum*, xxxviii, and in Bede's work of the same title, chapter xxxvi. Extracts from the former are found in MSS. Tiberius B.V, fols.

49*v*–51*r*, and Tiberius C. I, fols. 36*v*–38*r*; from the latter
in MS. Titus D. XXVI, fol. 5*r*–6*r* (printed by Wright and
Halliwell, *Reliquiae Antiquae*, i. 15, and more correctly by
Birch, *Liber Vitae*, p. 225). C. Wessely, in ' Bruchstücke einer antiken
Schrift über Wetterzeichen', *Wiener Sitzungsberichte*, cxlii (1900),
p. 37, discusses the most important ancient writings on meteorology
and the various possible approaches to the subject. G. Hellmann,
Denkmäler mittelalterlicher Meteorologie, Berlin, 1904, offers discussions
and reprints of several medieval tracts on meteorology.

13–14. Cp. Th. O. Wedel, ' The Mediæval Attitude Toward Astrology '
(New Haven, 1920), *Yale Studies in English*, lx. 24, 28 note 3 ; Philippson, op. cit., p. 111.

13–15. The passage from Byrhtferð quoted in the Parallels is taken
over literally from Ælfric, with two exceptions: (1) Instead of
Ælfric's *Hit is gecyndelic* Byrhtferð says *Ðā glēawe sǣgengan wel hig
understandað*. At first sight it is puzzling why 'wise mariners' are
supposed to know about the moon's influence upon terrestrial bodies.
However, this is merely a very bad blunder of Byrhtferð who mis-
understood Bede's *animantium cerebra maritimorum* (*DTR.* xxviii;
Giles, vi. 200, 2). A pretty effort in unconscious humour! (2) Byrhtferð 158, 8 adds the words *oððe fīf* when Ælfric says that the sea
always flows four 'points' later than it did the previous day. Crawford omits the two words from his translation without explaining
why. The reason for Byrhtferð's addition is his uncertainty as to
when an hour should be divided into four, and when into five 'parts'.
See the note on iii. 9 above.

15. The correspondence between the phases of the moon and the
tides is treated more fully by Bede, *DNR.* xxxix. In the *Glossae et
Scholia* in Migne's edition (*Patrologia Latina* xc. 423) of *DTR.* xxix
there is a diagram showing the relation of moon and tides. MSS.
Caligula A. XV, fol. 127*r*, Tiberius B. V, fol. 49*v*, and Tiberius C. I,
fol. 36*v* have Latin texts entitled *De Concordia Maris et Lunae*
which, however, are not taken from Bede. Different from these again is
the *Ordo lunae et decursus maris quando crescit et decrescit* in tħe Leofric
Missal (ed. Warren, Oxford, 1883, p. 57): it is a table, not a text. An
OE. text which calls itself the *endebyrdnes mōnan gonges and sǣflōdes*
is found in MS. Titus D. XXVII, fol. 56*v* (printed by Napier, *Anglia*,
xi. 6). It may ultimately derive from Bede, *DNR.* xxxix, although
there are certain discrepancies (cp. *Englische Studien*, lxix. 343).

CHAPTER IX

1–3. The 'mistaken popular belief' that stars fall from the heavens
was occasioned partly by the observation of meteors, 'shooting stars',
as Isidore, Bede, and Ælfric say, but partly also by the Biblical
prophecy that at the end of the world the stars will fall upon the earth:
Matt. xxiv. 29; Mark xiii. 25; Rev. vi. 13 and xii. 4; also Luke

xxi. 25–33. These prophecies are repeated in OE. literature: *Se Godspellere Lucas awrât on ðisum dægðerlican godspelle, þæt ure Drihten wæs sprecende þisum wordum to his leorning-cnihtum, be ðam tácnum ðe ǽr þyssere worulde geendunge gelimpað . . . Matheus se Godspellere awrât swutelicor þas tâcna, þus cweðende, 'þærrihte æfter ðære micclan gedrefednysse, bið seo sunne aðystrod, and se môna ne sylð nán leoht, and steorran feallað of heofonum.'* Ælfric, *Homilies*, i. 608–10. *He* (sc. *Antecrist*) *deð, þæt fyr cymð færlice ufan, swylce hit of heofonum cume, and egeslice forswælð fela þinga on eorðan.* Wulfstan, Homily xlii (ed. Napier, p. 195). A Latin text on falling stars is found in MS. Bodley 614 (twelfth century), fol. 36 v (cp. M. R. James, *Marvels of the East*, p. 8).

The idea that shooting stars betoken evil originated in ancient Babylonia (cp. Bruno Meissner, 'Babylonische Bestandteile in modernen Sagen und Gebräuchen', *Forschungen und Fortschritte*, 8. Jahrgang, No. 32) and gave rise to an astounding variety of beliefs throughout the Western world. It was known to the Greeks and Romans (cp. Wilhelm Kubitschek, 'Grundriss der antiken Zeitrechnung', *Handbuch der Altertumswissenschaft*, i. 7, München, 1928, p. 2) and recurs in the Vǫluspa where it is said that the stars will fall at the end of the world. The parallel conception that the moon will fall, or be eaten up by a monster, was equally widespread in classical and Germanic antiquity. It was caused by the eclipses of the moon (cp. Hoops's *Reallexikon der germanischen Altertumskunde*, iii. 236, s.v. *Monddämonen*). Ælfric fails to mention this belief in his discussion of the eclipse (*De Temporibus Anni*, iii. 6), which is rather surprising since he is always anxious to refute popular superstitions. Yet one would think that the idea must have been known in Anglo-Saxon England. Hrabanus Maurus in his *Homilia de festis* (Migne, cx. 76–80) reports that the German people of his day tried to assist the moon during an eclipse by shouting, lighting big fires, and shooting arrows (cp. Ehrismann, *Geschichte der deutschen Literatur bis zum Ausgang des Mittelalters*, München, 1918, i. 52). They encouraged the moon by repeating the cry of the ancient Romans: *vince luna* (Hoops's *Reallexikon*, loc. cit.). The OHG. Muspilli tells us that the moon will fall on Judgement Day.

Falling stars are known to this day in popular lore as harbingers of misfortune or of the end of the world (cp. *Handwörterbuch des deutschen Aberglaubens*, viii. 469–76, Berlin und Leipzig, 1936–7, s.v. *Sternschnuppe*). The idea was given expression in folk-song (cp. Otto Böckel, *Deutsche Volkslieder aus Oberhessen*, Marburg, 1885), and Clemens Brentano, the romanticist, incorporated this song in his tale of *Kasperl und Annerl:*

Wann der Jüngste Tag wird werden,
Dann fallen die Sternelein auf die Erden.

The poetic appeal of the idea was felt, too, by the seventeenth-century dramatist Andreas Gryphius (*Papinian*, ii. v. 295 sqq.):

Brecht Himmel! Sternen kracht! Sprützt schwefel-blaue Flammen!
Ihr Lichter jener Welt fällt! Klippen stürtzt zusammen
Und werft den Grund der hart befleckten Erden ein!
and by the modern Swiss poet Spitteler:
Allein am jüngsten Tage . . .
Dass Sonne, Mond und Sterne wie die Scharlachschuppen
Vom Himmel hageldicht zur Hölle schnuppen.

On the other hand, the pseudo-learned explanation of shooting stars
which Ælfric borrowed from Isidore also survived through many
centuries. It was given by Honorius Augustodunensis in his *Elucidarium sive dialogus de summa totius Christianae theologiae*, written
about 1092. (For a fragmentary OE. translation cp. Max Förster,
'Two Notes on Old English Dialogue Literature', *An English Miscellany presented to Dr. Furnivall*, Oxford, 1901, pp. 86–106.) The
Elucidarium was translated into many vernacular languages (Förster,
loc. cit.), the German version being known in a number of incunabula
(Hain, *Repertorium bibliographicum*, Nos. 8803 sqq.). On it is based
the account of the 'stars that fall upon the earth' in chapter 31 of the
chap-book of Dr. Faustus (cp. Robert Petsch's edition, Halle, 1911,
p. 185). This account agrees almost literally with Ælfric. (Cp. also
William Rose's modernization of the English Faust Book, London,
1925, p. 149.)

5. Translate: 'These seven are called the seven planets, and I know
that it will seem most unchristian (unbelieving) to unlearned men if
we discuss in detail *these* stars and their courses.' Cockayne in his
translation confused *ungelēaffullic* with *ungelēaflic*. Ælfric avoided a
translation of Bede's *DNR*. xii and xiii and put in its place a few
remarks about the Great Bear, the Pole Star, and the Pleiades (following Isidore). In *Anglia*, lviii. 292 sqq. I have shown that the
omission was due to Ælfric's fear lest he give food to the superstitious practices connected with the planets which were common in
his day. Later Ælfric overcame his scruples and in the *Interrogationes*,
xxi he gave an account of the planets based on the two chapters from
Bede which he had rejected in *De Temporibus Anni*. The insertion in
the *Interrogationes* is prefaced by the remark: 'I will say now what
I kept secret at an earlier occasion because of the evil practice of
astrology.' A further piece of evidence in favour of my interpretation, but which I overlooked when discussing the matter in *Anglia*,
lviii, is found in the following words which Ælfric added, going beyond
his source: . . . *þa seofon dweligendan tunglan. Hi synd dweliende oððe
woriende gecwedene, na for gedwylde, ac forþan þe heora ælc gæð on his
agenum ryne* (MacLean, ll. 117–19). Ælfric thought it necessary,
then, to explain that the term *dweliende tunglan*, 'roving stars' (translating Bede's *sidera errantia*), had nothing to do with *gedwyld*,
'heresy': which proves that in the popular mind it had! Byrhtferð did
not share Ælfric's misgivings. He translated Bede's *DNR*. xiii on
pp. 128, 19–130, 4 and his *DTR*. viii on p. 130, 5–30. The latter

chapter in Bede was also passed over by Ælfric. See the note on iii. 26 above.

8–9. A paraphrase, not altogether accurate, of these two paragraphs is found in John Lingard, *The History and Antiquities of the Anglo-Saxon Church*, London, 1845, ii. 171, note 2. The Pole Star is the last star in the tail of the constellation Ursa Minor, and now about 1° 2′ distant from the north pole. Isidore, in the passage which I have quoted as Ælfric's source, speaks of Boötes or Arctophylax, but it is difficult to say if he means α Boötis, α Ursæ Minoris, or one of these constellations themselves. The Pole Star was, and still is, the 'lodestar' of mariners. Its name, *stella maris*, was identified with the name of the Virgin, Maria (cp. note on i. 38–40 above). Thus she became the patron saint of seamen, and as such she was celebrated in Adam of St. Victor's (†1192) famous hymn:

Ave virgo singularis
Mater nostri salutaris
Quae vocatur stella maris
Stella non erratica.

The *Ave stella maris* was popular throughout the Middle Ages, and the theme was varied almost endlessly in Latin and vernacular religious song. (Cp. *Die Religion in Geschichte und Gegenwart*, vol. iii, s.v. *Mariendichtung*.) It was revived by the German Romanticists and lives to this day in folksong: *Meerstern ich Dich grüsse . . . Gib ein reines Leben, sichere Reis' daneben.*

10–12. The passage from the *Martyrologium* quoted in the Parallels is copied in MS. Harley 3271, fol. 92*v* (printed by me in *Englische Studien*, lxix. 347). The Pleiades were important because the ancients reckoned the beginning of summer and winter by their rise in the morning and evening respectively: *Horum autem principia temporum diuerse ponunt diuersi. Isidorus namque Hispaniensis episcopus, hyemem IX Calendarum Decemb. Ver, VIII Calend. Mart. Aestatem IX Calendas Junias, Autumnum X Calendas Septembres habere dixit exortum. Graeci autem et Romani, quorum in huiusmodi disciplina potius, quam Hispanorum auctoritas sequi consueuit, Hyemem VII Id. Nouemb. Ver. VII Id. Februa. Aestatem VII Id. Maii, Autumnum VII Id. Augusti inchoare decernunt, hyemis uidelicet et aestatis initia, uespertino uel matutino Vergiliarum ortu occasuque signantes. Item ueris et autumni, cum Pleiades media fere die uel nocte oriuntur et occidunt, ponentes ingressum.* Bede, *DTR.* xxxv (Giles, vi. 218). Cp. also Bede, *DT.* viii and *DTR.* xxx (Giles, vi. 205) ; Byrhtferð, p. 92 ; Piper, *Die Kalendarien und Martyrologien der Angelsachsen*, pp. 84 sqq. Isidore's dates for the beginnings of the seasons are copied in MS. Caligula A. XV, fol. 140*v*, and Bede's in the *Leofric Missal*, ed. Warren, p. 53. The latter dates were the more common by far in Anglo-Saxon England. They are given almost regularly in the calendars. See F. Wormald's edition *passim*.

13. The belief in comets as evil portents was general in Anglo-Saxon times. Cp. Philippson, *Germanisches Heidentum bei den Angelsachsen*, Leipzig, 1929, p. 224. The best-known example is, perhaps, the Bayeux Tapestry which shows King Harold viewing the comet that announces his own downfall. The Elucidarium and the chapbook of Dr. Faustus, cap. 28, again offer close parallels to Bede's and Ælfric's texts. See the note on ix. 1–3 above.

CHAPTER X .

1–13. The sources I have given for this passage are very inadequate. Either Ælfric had a subsidiary source which I have failed to discover (Isidore's *DNR*. xi is no nearer to him than Bede) or else he treated the discussion of the four elements in Bede's *DNR*. iii, iv, and xxv with unusual freedom. He was obviously unwilling, or even afraid, to repeat the time-honoured theory of the 'physical and physiological Fours', i.e. the supposed correspondence between the elements, the seasons, and the ages and humours of man, which is expounded in Bede's *DTR*. xxxv. Byrhtferð refers to it many times and has several diagrams illustrating it, but in Ælfric caution of this kind is not surprising. He may have felt that it bordered upon superstition to offer a glib equation between the macrocosm and man, the microcosm. Bede's system is this:

Spring, the air, childhood, sanguineness are warm and wet.
Summer, fire, youth, choler are warm and dry.
Autumn, the earth, manhood, melancholy are cold and dry.
Winter, water, old age, phlegm are cold and wet.

The elements themselves, then, are not strictly divided but mix or overlap in their qualities because (as Bede explains in *DNR*. iv) both earth and water are cold, water and air are damp, air and fire are warm and, finally, fire and earth are dry. It was this observation which Ælfric must have had in mind when he wrote paragraphs 9–11. Only he says that every body contains all four elements, and he preferred to give a concrete example rather than to repeat Bede's abstract system. A fuller example is found in the *Hexameron*, ll. 407–12: *Fýr is behyd on heardum stanum; se stan cymð of eorðan 7 he swæt swa ðeah, 7 of stancludum cumað wyllspringas. Ure lichama is eorðe 7 he oft ðeah swæt, 7 of ðam fyre hata ð, ðe him on wunað, 7 on ðam lyfte we lybbað ealle.* Ælfric's idea is reflected in the early modern English use of the verb 'to element' and the adjective 'elemented', meaning 'to compound (compounded) of the (four) elements'. Cp. *OED*. iv. 83 sq. John Donne's poem *Loves growth* describes love as 'elemented' or 'mixt of all stuffes' (ed. H. J. C. Grierson, London, 1933, p. 30 sq.).

17–23. Bede's chapter on the twelve winds is based on Isidore's *DNR*. xxxvii which, in turn, was taken from Suetonius' *Prata* (according to G. Hellmann, op. cit., pp. 14 and 39, note 9). Hellmann points out that it was through Isidore that the system of twelve winds, and twelve points of the compass, became the accepted one throughout the Middle Ages. He claims (ibid., p. 21) that Alcuin introduced the English and German names of these twelve winds, i.e. their description by combining the names of the four major points of the compass. The modern system, division of the compass into 8, 16, or 32 points, was first used by mariners, probably in the fourteenth century.

21–23. A Latin text on the four principal winds is found in MS. Bodley 614, fol. 34*v* (cp. James, *Marvels of the East*, p. 7), and a table showing these four winds in MS. Titus D. xxvii, fol. 21*v* (cp. Birch, *Liber Vitae*, p. 276). All twelve winds are entered in a scheme in the *Glossae et Scholia* appended to Bede's *DTR*. xxix (Migne, xc. 423–4), and they are enumerated in Bede's *DNR*. xxvii. A somewhat mutilated list is found in MS. Additional 32246, fol. 10*v*, and MS. Tiberius C. I, fol. 11*r*, has both a diagram and a list (in the margin) giving the winds, the latter badly distorted. The lists in Additional and Tiberius are glossed in Anglo-Saxon, and the diagram in the latter MS. also has a few glosses. I print below in five parallel columns: (1) The table of winds in the *Glossae et Scholia*. Its arrangement follows the points of the compass. I have rearranged the other four lists to agree with this order, but the numbers indicate the original order of each list. (2) Bede's list in *DNR*. xxvii. It begins with the north wind and always names first one of the four principal winds and then the two winds to the right and left of it. (3) The list in MS. Additional 32246, entitled *Nomina XII. Ventorum*. It is printed in Wright-Wülker, *Anglo-Saxon and Old English Vocabularies*, London 1884, column 143. The edition is unreliable because it is based on Junius' transcript, not on the original MS., as was shown by Max Förster in *Anglia*, xli. 94 sqq. To Professor Förster I owe the collations with MS. Additional. See the note on iii. 18–25 above. The list itself is incomplete, lacking the name of the north-east wind, Volturnus. The reason for the omission is found in the fact that the south-east wind is here named Volturnus. It is more commonly called Eurus, but this name is here identified with Euroauster. The OE. glossarist recognized the four cardinal winds but did not know the other eight. Thus four of his glosses (those on Eurus, Euroaffricus, Circius and Aquilo) are incorrect. In the original order this list enumerates the four principal winds first, and then follow in pairs the south-easterly, south-westerly, north-westerly, and north-easterly winds. The glossarist must have thought the list before him to be arranged in the manner of Bede's, which accounts for all his mistakes. (4) The names in the diagram of MS. Tiberius C. I. It is a circular scheme essentially the same as the one in the *Glossae et Scholia*. There are graphic variants in the Latin names and one error in the glosses: Euroauster or Euronothus (the scribe gave it two glosses, probably thinking that the names stood for two different winds) is glossed *sūðanwestan* whereas it should be *sūðanēastan*. (5) The list in MS.

Tiberius C. I. It is quite corrupt, giving only ten winds (three names are drawn into one) and having no intelligible system of arrangement. The scribe did not know Latin and concocted new names by combination with the Latin preposition *ab*: Abeuro, Afauonio, Acircio, Aborea, probably on the analogy of Africo (really Africus). In his source the names Eurus, &c., must have been used, not for the winds, but for the points of the compass. Only three words are correctly glossed (Subsolanus, Euroafricus, Chorus), very likely by accident. There follow in the MS. five more words with glosses which again show the scribe's ignorance or negligence: 'aboriente, *estan norðan*. a meridie *uel* ab austro, *estan*. ab occidente, *suðan*. A septentrione, *westan*. Ab aquilone, *norðan*.' It is evident that the glosses are copied incorrectly from the prototype, and not added by the scribe of Tiberius, because if they were moved up one place they would be in their proper order. Six more OE. glosses on the names of winds (two of them incorrect) may be found in A.S. Napier, *Old English Glosses*, Oxford, 1900, pp. 118, 146, 191, 194, 197 (two).

	Glossae et Scholia.	*DNR. xvii.*	*MS. Additional 32246.*	*MS. Tiberius C.I (diagram).*	*MS. Tiberius C.I (list).*
East	(1) Subsolanus	(4) Subsolanus qui et Apeliotes	(1) Subsolanus, easten wind	(1) Subsolanus, Apoliotes, est wind	(1) Subsolanus, estan
East-south	(2) Eurus qui et Vulturnus	(6) Eurus	(5) Vulturnus, easten suðan wind	(2) Eurus, estan suðan / Euroauster, suðan westan	(7) Eurus, suðan westan
South-east	(3) Euronotus qui et Euro-Auster	(8) Euroauster	(6) Eurus, euroauster, easten wind	(3) Euronothus, westan suþan	(5) Abeuro, westan suðan suðan
South	(4) Auster	(7) Auster qui et Notus	(2) Auster uel nothus, suðen wind	(4) Auster, nothus	(4) Nothus, suþan westan
South-west	(5) Libonotus qui et Auster-Africus	(9) Euronotus	(7) Euroaffricus, suðan easten wind	(5) Austerafricus, Libo-nothus	(6) Afauonio 7 africo 7 euro, suðan westan
West-south	(6) Africus qui et Libs	(11) Africus qui et Libs	(8) Affricus, suþan westan wind	(6) africus, suþan westan westan	(3) Africo, norþan westan
West	(7) Favonius	(10) Zephyrus qui et Favonius	(3) Fauonius uel zephirus, westen wind	(7) Fabonius, Zephirus, westan	(9) Zephirus, estan norþan
West-north	(8) Chorus Argestes	(12) Corus qui et Argestes	(9) Chorus, norðan westan wind	(8) Chorus, agrestis	(10) Chaurus, norðan westan
North-west	(9) Circius qui et Tracias	(2) Circius qui et Thrascias	(10) Circius, norðan easten wind	(9) Circius, trascias	(2) Acircio, suþan westan
North	(10) Septentrio	(1) Septentrio qui et Aparctias	(4) Septentrio, norþan wind	(10) Septentrio, Apartias	
North-east	(11) Aquilo qui et Boreas	(3) Aquilo qui et Boreas	(11) Aquilo uel boreas, norðan westan wind	(11) Aquilo, Boreas	(8) Aborea, estan suðan
East-north	(12) Vulturnus qui et Boetias	(5) Volturnus qui et Caecias		(12) Vulturnus, Calcias	

01 14

The manufacturer's authorised representative in the EU for product
safety is Oxford University Press España S.A. of El Parque Empresarial
San Fernando de Henares, Avenida de Castilla, 2 - 28830 Madrid
(www.oup.es/en or product.safety@oup.com). OUP España S.A. also acts
as importer into Spain of products made by the manufacturer.
Printed and bound by CPI Group (UK) Ltd, Croydon, CR0 4YY
16/12/2025
02020405-0002